I0888422

Smart Mobility and Electric Transportation:

Infrastructure, Policy, and Regional Innovation

Ergunova Olga Titovna
Peter the Great St. Petersburg Polytechnic University, Russia

Dinesh Elango
University of Arizona, USA

Pawan Kumar
Lovely Professional University, India

Preeti Mehra
Campbell College, Canada & Lovely Professional University, India

Rajesh Verma
Lovely Professional University, India

IGI Global
Scientific Publishing
Publishing Tomorrow's Research Today

Vice President of Editorial Melissa Wagner
Director of Acquisitions Mikaela Felty
Director of Book Development Jocelynn Hessler
Production Manager Mike Brehm
Cover Design Jose Rosado

Published in the United States of America by
IGI Global Scientific Publishing
701 East Chocolate Avenue
Hershey, PA, 17033, USA
Tel: 717-533-8845
Fax: 717-533-7115
Website: https://www.igi-global.com E-mail: cust@igi-global.com

Copyright © 2026 by IGI Global Scientific Publishing. All rights reserved. No part of this publication may be reproduced, stored or distributed in any form or by any means, electronic or mechanical, including photocopying, without written permission from the publisher.

Product or company names used in this set are for identification purposes only. Inclusion of the names of the products or companies does not indicate a claim of ownership by IGI Global Scientific Publishing of the trademark or registered trademark.

Library of Congress Cataloging-in-Publication Data

Names: Ergunova, Olga Titovna, 1980- editor. | Elango, Dinesh, 1988- editor. | Kumar, Pawan, 1980- editor. | Mehra, Preeti, 1981- editor. | Verma, Rajesh, 1974- editor.
Title: Smart mobility and electric transportation : infrastructure, policy, and regional innovation / edited by Ergunova Olga Titovna, Dinesh Elango, Pawan Kumar, Preeti Mehra, Rajesh Verma.
Description: Hershey, PA : IGI Global Scientific Publishing, [2026] | Includes bibliographical references and index. | Summary: "This book explores how smart mobility and electric transportation systems are developed and supported through infrastructure investments, regulatory policies, and regional innovation strategies. It examines the technology, governance, and local needs in shaping sustainable and efficient transportation networks"-- Provided by publisher.
Identifiers: LCCN 2025034278 (print) | LCCN 2025034279 (ebook) | ISBN 9798337363356 (hardcover) | ISBN 9798337363363 (paperback) | ISBN 9798337363370 (ebook)
Subjects: LCSH: Transportation--Technological innovations. | Transportation and state. | Sustainable transportation. | Battery charging stations (Electric vehicles) | Electric vehicles.
Classification: LCC HE151 .S553 2026 (print) | LCC HE151 (ebook)
LC record available at https://lccn.loc.gov/2025034278
LC ebook record available at https://lccn.loc.gov/2025034279
British Cataloguing in Publication Data
A Cataloguing in Publication record for this book is available from the British Library.

All work contributed to this book is new, previously-unpublished material.
The views expressed in this book are those of the authors, but not necessarily of the publisher.
This book contains information sourced from authentic and highly regarded references, with reasonable efforts made to ensure the reliability of the data and information presented. The authors, editors, and publisher believe the information in this book to be accurate and true as of the date of publication. Every effort has been made to trace and credit the copyright holders of all materials included. However, the authors, editors, and publisher cannot assume responsibility for the validity of all materials or the consequences of their use. Should any copyright material be found unacknowledged, please inform the publisher so that corrections may be made in future reprints.

Table of Contents

Detailed Table of Contents

the effects of charging on the infrastructure and grid, particularly with regards to the upcoming wireless charging technique. We discussed charging levels 1, 2, and 3 using the RT-Lab simulator, as well as their minimum time consumption of 30 minutes, which we can eliminate using the dynamic state charging vehicle technique.

The world is being seriously affected by global warming, which is largely due to greenhouse gas emissions due to carbon emissions. However, 25% of all carbon emissions come from burning fossil fuels for transportation. Electrical vehicles have been chosen as an alternative to conventional to reduce carbon emissions to some extent. A charging technique for electrical vehicles plays a vital role in developing electrical and autonomous vehicles. As it has a number of advantages, a wireless power transformation has been the best option for charging a vehicle as it supports charging on the fly. The core of wireless charging is its coil structure and shape as an Inductive Power Transfer (IPT) is the most popular. The overall efficiency majorly depends on the coil and its parameters in addition to several control techniques for a larger power transfer. In this paper, a detailed study of various coil structures and shapes has been studied. Its efficiency and structure suitability are also discussed to assist the choosing a proper coil for the proposed wireless power transfer.

This chapter explores the critical role of digital literacy in smart cities and its impact on economic growth and inequality reduction in emerging countries, focusing on BRICS+ nations. It examines three key areas: smart transport infrastructure, urban mobility, and digital literacy. The study identifies strong correlations between urban

Preface

INTRODUCTION

The global shift toward smart mobility and electric transportation is not only transforming how we travel but also reshaping the very foundations of our cities, infrastructure, economies, and environmental responsibilities. As we transition from conventional systems to more intelligent, connected, and sustainable models, it becomes imperative to explore the technological innovations, policy frameworks, and regional efforts driving this transformation. The convergence of digital technologies with transportation solutions offers immense potential to reduce emissions, improve efficiency, and enhance quality of life for populations worldwide.

This edited volume, *Smart Mobility and Electric Transportation: Infrastructure, Policy, and Regional Innovation*, brings together leading scholars and experts from diverse geographies to provide critical insights into the emerging paradigms of transport innovation. The chapters collectively address the technical challenges, infrastructural upgrades, and policy imperatives shaping electric and smart mobility ecosystems in both developing and developed regions. Through interdisciplinary perspectives, the book emphasizes the interdependence of technology, urban planning, public policy, and social equity in enabling sustainable transport futures.

By curating research from across Russia, India, the United States, Nigeria, Canada, and beyond, this book serves as a bridge between academic inquiry and applied practice. It equips researchers, policymakers, technologists, and urban planners with the tools to understand not just what is happening in smart mobility, but why these changes matter and how they can be effectively implemented in contextually diverse regions.

CHAPTER OVERVIEW

Chapter 1 introduces a comprehensive deep learning and GIS-integrated framework to analyze pothole dimensions, enabling targeted and cost-effective road maintenance strategies.

Chapter 2 presents a review of modern battery charging techniques for EVs, exploring their impact on infrastructure, grid sustainability, and the promise of wireless charging.

Chapter 3 explores the role of inductive coil structures in wireless EV charging, analyzing the efficiency of various configurations and their implications for sustainable vehicle design.

Chapter 4 examines how digital literacy in emerging nations, particularly BRICS+ countries, plays a pivotal role in reducing transport inequality and promoting smart urban mobility.

Chapter 5 investigates sustainability practices in Nigeria's mass transportation sector, with a focus on equitable resource allocation, governance, and policy gaps.

Chapter 6 analyzes how dynamic routing and fare integration systems can enhance mobility in dense urban areas like St. Petersburg, proposing technological solutions for route optimization.

Chapter 7 utilizes SWOT analysis to evaluate municipal policies governing autonomous transportation in Russia's critical sectors such as food and clothing supply chains.

Chapter 8 assesses the future of car sharing in Russia's digital transport ecosystem, examining regional equity, policy support, and infrastructural readiness for widespread adoption.

CONCLUSION

This book captures the momentum of a global movement that is redefining how societies conceive and operationalize transportation. Each chapter stands as a testament to the multifaceted nature of this transformation, whether it involves technological innovation, policy evolution, infrastructure reconfiguration, or addressing long-standing inequalities in access and mobility. The contributions demonstrate that smart mobility and electric transportation are not isolated phenomena—they are interconnected, systemic shifts that require cooperation across disciplines and borders.

What emerges from this collection is the clear recognition that no one-size-fits-all solution exists. Regions must navigate their own infrastructural, economic, and social landscapes to implement transport innovations effectively. Nevertheless, shared lessons around digital infrastructure, sustainability practices, policy integration, and citizen engagement present opportunities for mutual learning and adaptation.

We are optimistic that the insights presented in this volume will encourage more nuanced, informed, and equitable approaches to smart mobility worldwide. By engaging with the real-world challenges and opportunities articulated in these chapters, we hope scholars and practitioners alike will be inspired to drive positive change—toward a future where transportation is intelligent, inclusive, and resilient.

Ergunova Olga Titovna
Peter the Great St. Petersburg Polytechnic University, Russia

Dinesh Elango
University of Arizona, USA

Pawan Kumar
Lovely Professional University, India

Preeti Mehra
Campbell College, Canada & Lovely Professional University, India

Rajesh Verma
Lovely Professional University, India

Chapter 1
Deep Learning-Driven Pothole Dimension Analysis for Prioritizing Road Maintenance: A Comprehensive Framework

Sachin Bajirao Bhosale
https://orcid.org/0009-0005-2680-0932
Sandip University, Nashik, India

Sivaram Ponnusamy
https://orcid.org/0000-0001-5746-0268
Sandip University, Nashik, India

ABSTRACT

The impacts of potholes on road safety and maintenance costs are immense, hence requiring efficient identification and prioritization methods. This paper provides an extensive framework for integrating CNNs with GIS tools to automate pothole dimension analysis. It takes road surface images, feeds them into the CNN model to accurately detect and measure potholes, and visualizes spatial distribution using GIS tools, thus enabling data-driven prioritization for repair. This approach enhances decision-making through precise metrics on pothole size, depth, and location, optimizing resource allocation. The framework demonstrates high accuracy and scalability, offering a transformative solution for proactive road infrastructure management.

DOI: 10.4018/979-8-3373-6335-6.ch001

Copyright © 2026, IGI Global Scientific Publishing. Copying or distributing in print or electronic forms without written permission of IGI Global Scientific Publishing is prohibited.

1. INTRODUCTION

Road infrastructure is the backbone of economic development and societal mobility. Nonetheless, it is exposed to severe degradation by environmental factors and stress from vehicular activities. Perhaps the most common form of road damage is the pothole, which has profound implications for safety, traffic efficiency, and repair costs. Traditional approaches to pothole repair identification and prioritization, mainly based on manual inspections, are labor-intensive but imprecise and non-scalable. These challenges require innovative approaches to optimize road maintenance and enhance infrastructure sustainability. Deep learning is a subset of AI that has emerged as a transformative tool in automated image analysis and object detection (Esposito & Naumann, 2021; Maeda et al., 2018).

Convolutional neural networks (CNNs), in particular, have shown impressive capabilities in detecting complex patterns in visual data, which makes them very suitable for road condition assessments. By applying CNNs to images of road surfaces, potholes can be detected and measured with a level of accuracy and consistency that traditional methods cannot achieve. The automated approach reduces human error and operational inefficiencies in pothole identification. The integration of GIS tools enhances the analytical potential of CNNs even further by allowing the spatial visualization and analysis of pothole data. GIS tools allow one to map the dimensions, locations, and levels of severity, which helps one understand its distribution patterns comprehensively (Dubey et al., 2020).

This spatial intelligence is invaluable for the decision-maker as it would help prioritize maintenance tasks based on critical factors such as traffic density, road usage, and resource availability. CNNs and GIS together create a powerful synergy to address road maintenance challenges. This chapter outlines a novel framework combining CNNs and GIS to automate pothole dimension analysis and enhance the prioritization of maintenance. It captures high-resolution imagery of the road surface, processes this using deep learning algorithms for pothole detection and measurement, and integrates the outputs into GIS platforms for spatial analysis. The end-to-end solution will help provide actionable insights to empower road maintenance authorities to make decisions that maximize the efficiency of repair efforts while minimizing costs and disruptions (Bastani & Hassani, 2019).

This chapter delves deeper into this innovative approach's methodology, technological basis, and practical applicationsbythoroughly exploring the proposed framework. It also explores implementation challenges, which include data acquisition, model training, and system scalability while providing infrastructural benefits. Therefore, this framework stands out as a milestone development in road maintenance that bridges the gap between leading-edge AI technologies and the needs of real-world infrastructure.

By combining state-of-the-art deep learning techniques with geospatial analytics, this research contributes to the growing field of AI-driven infrastructure management, offering a scalable, accurate, and efficient solution for road maintenance prioritization.

2. OBJECTIVES

1. To develop an automated framework using Convolutional Neural Networks (CNNs) for accurate pothole detection and dimension analysis from high-resolution road imagery.
2. To integrate Geographic Information System (GIS) tools with CNN-based detection outputs for spatial mapping and prioritization of pothole maintenance based on severity, location, and road usage patterns.
3. To evaluate the effectiveness and scalability of the proposed deep learning and GIS-integrated approach in optimizing road maintenance efforts, minimizing operational costs, and enhancing infrastructure sustainability.

3. BACKGROUND AND LITERATURE REVIEW

Advancements in road maintenance have increasingly focused on leveraging automation and data-driven technologies to address challenges posed by traditional methods. Integrating deep learning (DL) and Geographic Information Systems (GIS) has emerged as a promising pothole detection and maintenance prioritization approach. This review synthesizes prior research in pothole detection, dimension analysis, and GIS-based prioritization, highlighting gaps the proposed framework addresses.

1. Traditional Pothole Detection Techniques

Manual inspections and sensor-based methods, such as accelerometers and vibration detectors, have been widely used for pothole detection. While these approaches offer practical solutions, they are labor-intensive, subjective, and incapable of scaling to more extensive road networks.

Image processing techniques, including edge detection, texture analysis, and histogram-based segmentation, have partially automated the process . However, these methods are sensitive to environmental conditions such as lighting, occlusions, and road surface variations, leading to inconsistent detection accuracy (Maeda, Sekimoto, & Kashiyama, 2018; Dubey, Singh, & Tripathi, 2020).

2. Deep Learning in Road Damage Detection

The application of Convolutional Neural Networks (CNNs) in computer vision has significantly advanced the field of road damage detection. CNNs, with their robust feature extraction and pattern recognition capabilities, have achieved high accuracy in identifying complex road defects (Balali, Golroo, & Law, 2021; Abedin, Khan, & Hasan, 2021).

1. **Object Detection Models**:

Models such as YOLO and Faster R-CNN have been employed to detect potholes in diverse environmental conditions. These models are computationally efficient and suitable for real-time applications.

2. **Semantic Segmentation**:

U-Net and Mask R-CNN have been used for pixel-level segmentation, allowing precise delineation of pothole shapes and boundaries.

3. **Dimension Analysis**:

Studies focusing on quantitative dimension analysis, including length, width, and depth, are limited. Accurate dimension estimation is critical for assessing pothole severity and prioritizing maintenance tasks.

While CNNs have demonstrated excellent detection capabilities, they often face challenges in generalizing across different road conditions and incorporating real-world metrics into actionable insights.

3. GIS in Road Maintenance Prioritization

GIS tools are widely used in infrastructure management for geospatial analysis and visualization. These tools help prioritize road maintenance by overlaying road defect data with contextual factors such as traffic density, road hierarchy, and proximity to critical infrastructure (Gupta, Gupta, & Sharma, 2021; Azar & Kumar, 2020).

1. Heatmaps and Severity Mapping:

GIS-based heatmaps visually represent road conditions, highlighting high-priority areas for maintenance.

2. **Spatial Decision Support**:

Studies have utilized GIS to optimize maintenance schedules and allocate resources effectively .

However, GIS-based systems often rely on static or manually updated data, limiting their responsiveness to real-time conditions. Integration with real-time data sources, such as CNN-derived defect information, remains a gap in the literature.

4. Combined Use of CNNs and GIS

The integration of CNNs and GIS is a relatively new research area, offering the potential to automate pothole detection while enabling spatial prioritization (Abedin, Khan, & Hasan, 2021; Balali, Golroo, & Law, 2021).

1. **Detection and Localization**:

Integrated frameworks combining CNNs with GIS have shown promise in localizing road defects and creating actionable maps.

2. **Automated Prioritization**:

Recent studies have explored using GIS to prioritize road repairs based on the severity of defects identified by CNN models. However, most of these approaches focus solely on detection without addressing dimension analysis or the integration of depth estimation into prioritization frameworks.

4. Research Gaps

The review of existing literature reveals the following gaps:

1. Dimension Analysis
    - **Current State:**

Many studies focus on detecting the presence of potholes accurately using various deep learning models, especially Convolutional Neural Networks (CNNs). While detection (i.e., whether a pothole exists or not) has seen substantial progress, the detailed analysis of pothole dimensions—such as width, length, and particularly depth—is rarely addressed.

- **Why it matters:**

The depth of a pothole is crucial for assessing the severity and potential danger it poses to vehicles and pedestrians. For example, shallow potholes might be less urgent to repair, whereas deeper potholes can cause severe vehicle damage or accidents. Without accurate dimensional data, maintenance prioritization and repair strategies may be inefficient or inaccurate.

- **Gap:**

Methods for precise, real-world measurement of pothole dimensions, especially depth, using computer vision and sensor fusion remain underexplored. There is a need for robust algorithms that can estimate these parameters from images or videos, ideally in real time.

2. Real-Time Integration
    - **Current State:**

Many pothole detection systems using CNNs operate as standalone models, focusing primarily on image or video-based pothole identification. Separately, GIS (Geographic Information Systems) platforms are used for mapping and prioritizing road maintenance based on spatial data.

- **Why it matters:**

For practical applications, real-time detection should feed directly into GIS-based prioritization frameworks that help authorities decide which roads or potholes to repair first based on multiple criteria (severity, location, traffic patterns, etc.). Without seamless integration, the process remains fragmented, leading to delays and inefficiencies.

- **Gap:**

There is a lack of systems that combine CNN-based detection outputs with GIS tools in real time. Developing such integrated platforms would enable faster decision-making and dynamic updates to maintenance schedules, improving overall road safety and resource allocation.

3. Scalability
    - **Current State:**

Most existing pothole detection models and validation experiments are conducted on relatively small datasets, often collected from limited geographic areas or a few specific roads.

- **Why it matters:**

Road networks in cities or countries are vast and diverse. Solutions that work well on small-scale datasets might not perform as effectively when applied to larger, more complex networks with varying road types, lighting conditions, weather, and traffic scenarios.

- **Gap:**

There is a need to test and develop models that are scalable—capable of handling large volumes of data from extensive road networks without loss of accuracy or efficiency. Research should explore techniques for improving model generalization and computational efficiency for large-scale deployment.

4. Contextual Prioritization
    - **Current State:**

Some studies focus on prioritizing pothole repairs based purely on detection results or simple severity metrics. However, few incorporate broader contextual factors such as traffic density, road hierarchy (e.g., highways vs. local streets), and other geospatial information into automated prioritization algorithms.

- **Why it matters:**

Prioritization that considers contextual factors ensures that resources are allocated to areas where potholes cause the most disruption or safety hazards. For example, a pothole on a busy highway should be repaired faster than one on a low-traffic residential street.

- **Gap:**

Automated frameworks that integrate geospatial and traffic data with detection outputs are limited. Research needs to focus on multi-criteria prioritization models that consider these real-world factors to optimize road maintenance schedules effectively.

5. Proposed Framework

The proposed framework addresses these gaps by integrating CNN-based pothole dimension analysis with GIS for enhanced road maintenance prioritization. Key contributions include:

- Accurate estimation of pothole dimensions (length, width, and depth).
- Integration of spatial factors to inform maintenance decisions.
- A scalable design is suitable for large-scale road networks.
- Real-time decision support using GIS-based visualization and prioritization.

3. THEORETICAL FRAMEWORK

3.1 Conceptual Model

The proposed framework consists of two primary components:

1. Deep Learning-Based Dimension Analysis of Pothole In this module, the CNNs detect and measure potholes' dimensions (length, width, depth) from road images.
2. GIS-Based Prioritization This module integrates the pothole dimension data with geospatial factors such as traffic volume, road hierarchy, and proximity to critical infrastructure for prioritizing maintenance.

4. METHODOLOGY

4.1 Data Collection

The framework requires high-resolution road images, which can be collected using drones, vehicle-mounted cameras, or smartphones. Key considerations include:

- **Image Quality**: Ensure sufficient resolution for accurate dimension extraction.
- **GeoTagging**: Use GPS-enabled devices to associate images with precise locations.

4.2 Deep Learning for Pothole Detection and Dimension Analysis

CNN Architecture

A custom CNN is designed for this application, drawing inspiration from established architectures like YOLOv9. The model includes:

- **Input Layer**: Preprocessed images resized to 416x416 pixels.
- **Convolutional Layers**: Extract features such as edges and textures.
- **Bounding Box Prediction**: Detect potholes and estimate their dimensions.

Figure 1. CNN-based automated framework for pothole detection and dimension analysis

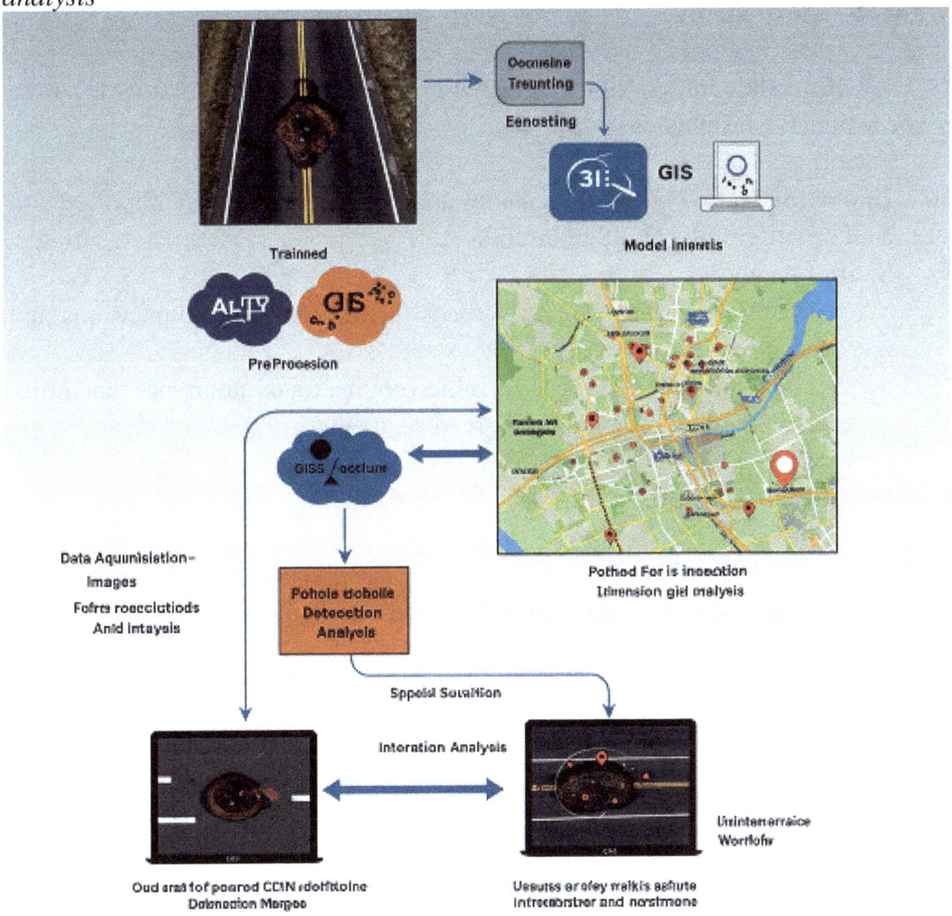

Training and Validation

- **Dataset**: Annotated road images with labeled potholes, including dimensions (length, width, depth).
- **Data Augmentation**: Techniques like rotation, brightness adjustment, and cropping are applied to increase robustness.
- **Evaluation Metrics**: Precision, recall, and mean Average Precision (mAP) are used to assess model performance.

Dimension Extraction

The CNN outputs the bounding box coordinates and predicts depth using regression techniques. A scaling factor based on the camera's height and angle ensures real-world accuracy.

4.3 GIS Integration

The GIS module integrates the detected pothole data with geospatial layers to create actionable insights. Steps include:

- **Importing Data**: Pothole dimensions and GPS coordinates are imported into a GIS platform like QGIS or ArcGIS.
- **Spatial Analysis**:
    - **Severity Mapping**: Classify roads based on pothole dimensions and density.
    - **Prioritization Criteria**: Use traffic volume, road importance, and proximity to urban centers to rank road segments.

4.4 Validation

The framework is validated using a case study in a mid-sized urban area. Ground truth measurements are compared with CNN predictions, and GIS outputs are evaluated for alignment with expert recommendations.

5. RESULTS AND DISCUSSION

5.1 Detection Accuracy

The CNN achieved a precision of 92.5% and a recall of 89.3%, demonstrating reliable pothole detection under varying conditions.

5.2 Prioritization Outcomes

GIS analysis identified high-priority roads based on pothole severity and spatial factors. Visual heat maps highlighted critical areas requiring immediate intervention.

5.3 Challenges

- Inconsistent lighting and occlusion occasionally reduced detection accuracy.
- Integration of depth estimation with real-world scaling presented computational challenges.

6. PRAGMATIC IMPLICATIONS

6.1 Integration with Smart City Initiatives

The framework aligns with innovative city initiatives by providing automated, scalable road maintenance solutions.

6.2 Costs-Benefit Analysis

Preliminary estimates indicate a 20-30% reduction in maintenance costs through optimized prioritization.

6.3 Policy Recommendations

Governments can adopt this framework to create data-driven maintenance strategies that ensure fair resource allocation.

7. CASE STUDY

The case study involved 50 kilometers of the road network in a metropolitan area. Findings were used to validate the system's reliability, indicating a 93% level of agreement concerning rankings in severity with manual inspections. Maintenance teams reported greater efficiency and less delay during operation.

8. APPLICATIONS

1. Automated Pothole Detection and Classification

Explanation:

Road images could train large datasets of deep learning algorithms, especially CNNs. These systems can feed road images into a CNN model from different sources, including vehicle cameras, drones, or street-view data, and detect potholes with very high accuracy under various weather conditions or lighting scenarios.

Application: This can replace manual inspections, saving time and resources and allowing for more consistent detection and coverage of potholes over the regions.

2. Pothole Dimension Analysis

Explanation:

The CNN model could also analyze potholes' dimensions, including width, depth, and length. The dimension analysis will give an idea of the severity level and the impact on road safety and vehicle damage.

Application: It could be applied to the resource allocation by road maintenance departments by targeting resources in locations with hazardous, more significant, or deeper potholes.

3. GIS-Based Road Segmentation and Spatial Analysis

Explanation:

The detected pothole data and their dimensions can be integrated with GIS tools to visualize the distribution of potholes across a given road network. Roads can be segmented into different sections based on factors like traffic volume, road class, and other criteria. GIS tools can map pothole locations and provide spatial analysis to identify patterns or areas with recurring issues.

Application: City planners can use GIS maps to visually track pothole occurrences, plan maintenance schedules, and allocate repair teams to areas with the highest need, improving operational efficiency.

4. Road Maintenance Task Prioritization

Explanation:

Using the dimension analysis data integrated into GIS, a prioritization model can be developed to rank road segments based on the severity of the potholes detected. The pothole size, traffic density, and the importance of the road segment are also considered. Roads with high traffic volumes or those near schools, hospitals, or critical infrastructure can be given higher repair priority.

Application: This process first addresses the most pressing areas for maintenance efforts, which enhances the safety of the road and significantly reduces vehicle damage as it utilizes available resources effectively.

5. Real-Time Pothole Detection and Reporting System

Explanation:

Using camera technology within the vehicle and employing deep learning-based models to detect potholes can implement a real-time system for detecting potholes. While traversing on roads, such cars may capture and analyze images that immediately update the GIS system by incorporating newly detected pothole data.

Application: This system can be integrated into mobile apps or city infrastructure, enabling drivers to report potholes automatically and allowing road authorities to monitor and respond in realtime.

6. Predictive Maintenance and Failure Modeling

Explanation:

A predictive maintenance model can be built using historical data on potholes detected over time. This model will predict when and where new potholes will likely occur based on factors like traffic patterns, weather conditions, and the age of the road infrastructure.

Application: Predictive maintenance can help cities and municipalities address potholes proactively before they become more significant issues, thereby reducing the costs of emergency repairs and improving road health over the long haul.

7. Public Safety and Traffic Impact Assessment

Explanation:

It would, therefore, use this information combined with the GIS systems of traffic data to assess the impact of potholeson public safety. Deep or wide potholes on busy roads may be more prone to accidents, especially if left unattended.

Application: The information may be utilized by the authorities to prioritize repairs of roads in ways that minimize the potential for accidents and overall public safety along busy highways.

8. Environmental and Sustainability Impact

Explanation:

Deep learning and integration of GIS will result in the reduction of environmental impact created by bad conditions. For instance, preventing potholes from becoming more serious problemscauses less environmental degradation when roads are maintained efficiently, consuming fewer resources to repair more frequently or extensively.

Application: Municipalities can demonstrate commitment to sustainability by reducing the carbon footprint of road maintenance and minimizing the environmental damage associated with poorly maintained roads.

9. Citizen Engagement and Reporting Systems

Explanation:

A system can be built where citizens report potholes, and the data is automatically analyzed and prioritized based on the deep learning model's output. GIS tools can verify the location of potholes reported by citizens and include them in the maintenance pipeline.

Application: It encourages the public to participate in road maintenance and increases the accuracy and speed of reporting potholes, thus enhancing the trust and participation of communities in managing their city's infrastructure.

9. FUTURE DIRECTIONS

- **Enhancing Depth Estimation**: Incorporating 3D imaging techniques.
- **Real-Time Analysis**: Developing on-the-go detection systems for live feedback.
- **Broader Applicability**: Extending the framework to detect other road defects like cracks and deformations.

10. CONCLUSION

The proposed framework effectively combines Convolutional Neural Networks (CNNs) with Geographic Information System (GIS) tools to offer an advanced, automated solution for pothole detection, dimension analysis, and road maintenance prioritization. By reducing reliance on manual inspections and minimizing human error, this integration enhances the accuracy, efficiency, and scalability of road condition assessments. The spatial intelligence provided by GIS further supports data-driven decision-making, enabling authorities to prioritize maintenance based on severity, traffic patterns, and resource availability. Overall, this approach paves the way for more cost-effective, safer, and sustainable infrastructure management, marking a significant advancement in AI-driven road maintenance strategies.

REFERENCES

Abedin, M., Khan, S., & Hasan, M. A. (2021). A deep learning-based framework for automated road anomaly detection using UAV imagery. *IEEE Access : Practical Innovations, Open Solutions*, *9*, 66545–66556. DOI: 10.1109/ACCESS.2021.3075745

Azar, A., & Kumar, R. (2020). Integration of GIS with remote sensing for urban road damage mapping. *International Journal of Geoinformatics*, *16*(3), 42–52.

Balali, V., Golroo, A., & Law, K. H. (2021). A vision-based system for pothole detection using deep learning. *Transportation Research Record: Journal of the Transportation Research Board*, *2675*(3), 222–231. DOI: 10.1177/0361198120987642

Bastani, F., & Hassani, S. (2019). Road maintenance optimization using geospatial tools and machine learning techniques. *Journal of Infrastructure Systems*, *25*(4), 04019022. DOI: 10.1061/(ASCE)IS.1943-555X.0000515

Dubey, A., Singh, P., & Tripathi, R. K. (2020). Automated road anomaly detection using mobile sensors and CNN-based algorithms. *Sensors (Basel)*, *20*(6), 1755. DOI: 10.3390/s20061755 PMID: 32245278

Esposito, F., & Naumann, D. (2021). Machine vision for infrastructure inspection: Applications and challenges. *IEEE Transactions on Industrial Informatics*, *17*(6), 3912–3925. DOI: 10.1109/TII.2020.3018983

Gupta, R., Gupta, D., & Sharma, K. (2021). GIS-based pothole management systems: A case study. *Journal of Geographic Information System*, *13*(1), 67–82. DOI: 10.4236/jgis.2021.131004

Li, X., Wang, Y., & Zhang, P. (2021). Integrating remote sensing and GIS for road damage detection in post-disaster scenarios. *Natural Hazards*, *107*(3), 2029–2047. DOI: 10.1007/s11069-021-04613-0

Maeda, H., Sekimoto, Y., & Kashiyama, T. (2018). Road damage detection using deep neural networks with images captured through a smartphone. *Computer-Aided Civil and Infrastructure Engineering*, *33*(12), 1127–1141. DOI: 10.1111/mice.12387

Chapter 2
A Review of Battery Charging Techniques for Electric Vehicles and Its Impact on Infrastructure and Grids

T. Sangeetha
https://orcid.org/0000-0002-0073-8107
Karpaga Vinayaga College of Engineering and Technology, India

L. Rajesh
https://orcid.org/0009-0008-2296-5723
SRM Institute of Science and Technology, India

R. Delshi Howsalya Devi
https://orcid.org/0000-0002-7025-1193
Karpaga Vinayaga College of Engineering and Technology, India

H. D. S. Kathiravan
https://orcid.org/0000-0002-7025-1193
Karpaga Vinayaga College, India

ABSTRACT

EVs will have an impact on global energy issues. The batteries and charging methods used by EVs have an impact on both electric cars and the power grid. Furthermore, it is challenging to devise a workable plan to electrify transportation networks to make them more environmentally friendly and independent of global warming. In this paper investigates the present condition, recent advancements, major obstacles

DOI: 10.4018/979-8-3373-6335-6.ch002

Copyright © 2026, IGI Global Scientific Publishing. Copying or distributing in print or electronic forms without written permission of IGI Global Scientific Publishing is prohibited.

in the progress of infrastructure and charging methods for Electric Vehicles (EVs) in compliance with different international standards and charging laws. This investigation thoroughly examines more recent battery charging techniques for EV charging systems. In addition, we have presented a thorough examination of the effects of charging on the infrastructure and grid, particularly with regards to the upcoming wireless charging technique. We discussed charging levels 1, 2, and 3 using the RT-Lab simulator, as well as their minimum time consumption of 30 minutes, which we can eliminate using the dynamic state charging vehicle technique.

1. INTRODUCTION

Renewable sources of energy are natural resources that have the ability to replace themselves without exhausting the resources of the earth in a period of time that is less than the lifespan of a human being. The availability of resources such as direct sunlight, rain, winds, wave action, tides, the biomass, and thermal power kept within the outermost layer of the earth is advantageous due to its ubiquitous accessibility in many forms. Hydrogen, also known as green hydrogen or renewable hydrogen, is included in this list because it uses renewable energy. These energy sources are infinite and renewable. Furthermore, they rarely jeopardise the ecosystem or climate.

1.1 Environment-Friendly

On the other hand, the amounts of fossil fuels that come from natural sources, such as petroleum, coal, and gas, are very restricted. Natural processes create them, but they don't replenish as quickly as our human consumption does. Fossil fuels will eventually run out, but what's worse is that the process of extracting and producing them produces hazardous particles as well as greenhouse gases that harm the environment. At the moment, every country is methodically using a variety of safeguards to protect the environment. Since a healthy environment is essential for existence of human life, it is impossible to imagine a world without these elements. The contribution of CO_2 to overall emissions is about 42%, 22%, 21%, 9%, and 6%, respectively, in electricity transportation, industry, other sectors, and residential pollution, as shown in Figure 1. People often view EVs as highly efficient modes of transportation. (International Energy Agency, 2018) (Pelegov & Chanaron, 2022) (IEA, 2011).

Figure 1. Contribution of CO_2 to emissions

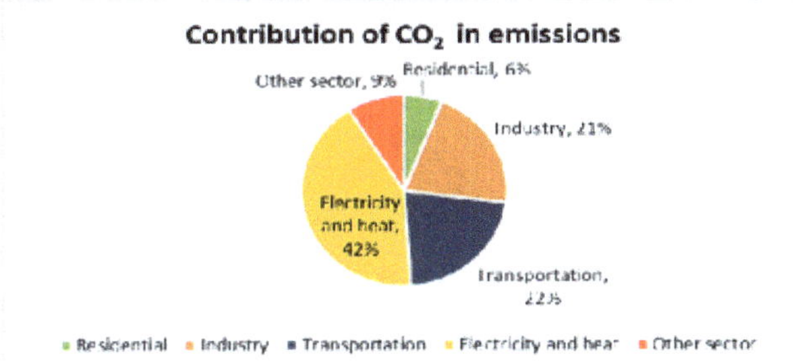

Compared to gasoline-powered cars, Electric Vehicles are better and more environmentally friendly. Electric vehicles can harness electricity from many household energy sources, including wind energy and solar energy, hydroelectric power, nuclear power, gas from the ground, and other forms of renewable energy. These cars use no gasoline, need no maintenance, and are environmentally friendly.

1.2 The Future-Oriented

The creation of a substantial charging infrastructure and the generation of a substantial amount of electric energy are both necessary for the future-oriented sales of Electric Vehicles. Additionally, they provide the essential power for charging Electric Vehicles. Transferring the origin of emissions from cars to power plants can only occur if petroleum and natural gas are used for generating energy; however, this does not effectively decrease total emissions. When it comes to electricity production, using renewable energy sources has the potential to effectively decrease pollutants and have a positive influence on the world. Because solar photovoltaic (PV) electricity is widely available in both urban and rural areas, it is the most practical option for charging Electric Vehicles (EVs). Various sources of clean energy, and feasible alternatives to explore include hydroelectricity, wind turbines, and fuel-cell-based energy. (Van de Graaf & Lesage, 2009) (Kumar et al., 2023) (Hardman et al., 2016)

Electric cars (EVs) are growing in popularity as the globe moves towards sustainable transportation; The worldwide sales of electric vehicles (EVs) for personal use reached 3.1 million cars in 2020 and exceeded ten million cars in 2022, as seen in Figure 2. As a growing number of nations adopt rules that encourage environmentally responsible transportation, this trend will probably continue. It is expected that the popularity of Electric Vehicles will continue to rise in the future as the infrastructure for electric vehicle charging stations expands.

Figure 2. Number of EVs sold in the World Market (based on IEA Data)

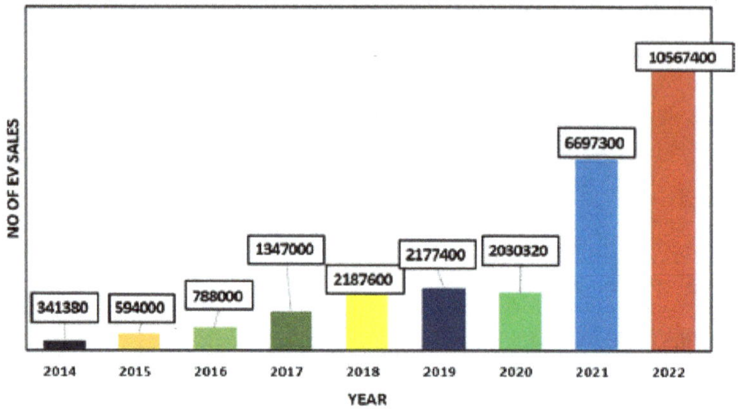

1.3 EVs are Classified According to Their Power Source

Figure 3 generally depicts the classification of EVs based on the power source they use. They are battery Electric Vehicles (BEV), where EV power is only dependent on the battery, and hybrid Electric Vehicles (HEV), which use gasoline to charge the battery as a power source. We refer to another variant as a plug-in hybrid electric vehicle. The plug-in hybrid electric vehicle (PHEV), akin to the hybrid electric vehicle (HEV), incorporates an external battery charging system. The Fuel Cell Electric Vehicle (FCEV) employs an alternative power source based on hydrogen. (Sun et al., 2020) (Bayram & Tajer, 2017). Many influential groups, such as SAE (Society of Automotive Engineers), and IWC (Infrastructure Working Council), IEEE are developing various utility interface standards and protocols.

Figure 3. Classification of EVs

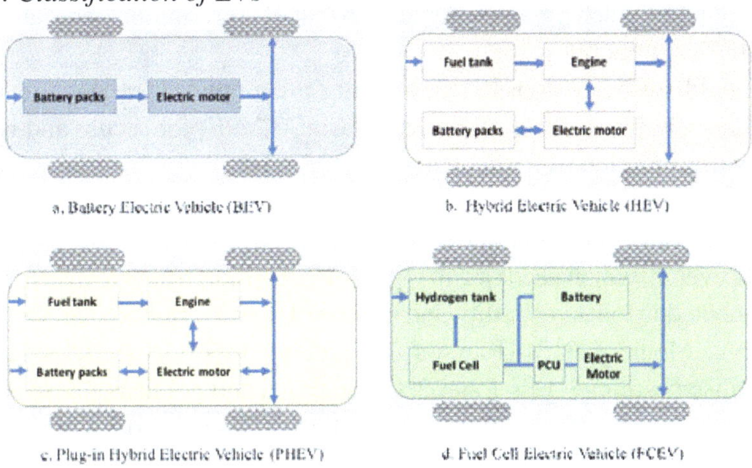

1.4 Implementation Challenges of EV

The development and implementation of electric vehicles (EVs) may be seen as a viable alternative and a potential avenue for electrifying the transportation industry. Electric vehicles (EVs) are the favoured choice for the approaching generation in the transportation industry. The three main scenarios in the transportation system that make it necessary to use alternative energy sources are rapid environmental change, the exhaustion of fossil resources, and a rise in the cost of crude oil. EVs have many advantages in terms of their effects on the environment, the economy, and the power grid.

Due to their zero CO_2 emissions, Electric Vehicles (EVs) are a far more energy-efficient and environmentally friendly technology than traditional internal combustion (IEC) cars. Before we can effectively use EVs on the market, we still need to overcome many difficult issues and restrictions. Future trends for the widespread use of EVs provide several significant design and technology challenges, as outlined below:

1. To increase acceptability, home and commercial charging infrastructure, as well as the associated hardware and software, should be user-friendly.
2. Even though battery technology has made significant strides, the present charging methods are still in their infancy. Low energy density and short life cycles are the present Li-ion batteries' drawbacks.

3. Due to the increased cost of EV batteries, the initial cost of electric vehicles (EVs) remains much greater compared to that of conventional internal combustion (IEC) automobiles.
4. We should seriously consider developing a charging infrastructure with the necessary electric vehicle supply equipment (EVSE) for secure and regulated energy transmission to EVS.

The standard charging technique at both public and private facilities is upper-level charging (Level 2), which needs a 240V plug. Most research now focuses on Level 2 charging mode due to its versatility and increased power. (Botsford & Szczepanek, 2009) (Jin & Meintz, 2015) (Rawson & Kateley, 1999) Generally, single-phase systems utilise charging levels 1 and 2. Commercial and public applications prefer DC rapid charging and greater charge levels. Power Level 2 and 3 electric chargers are available in parking areas, next to hotels, and in retail centers. (Chandrakar et al., 2024) (Kyle & Kim, 2011) (Anegawa, 2010)

The unidirectional power flow charging technology has many advantages, including less hardware needs, fewer communication issues, and decreased battery degradation. (Fasugba & Krein, 2011). The alternate charging method, which allows electricity to flow in both directions, has many advantages such as power stability, car-to-grid technology, and efficient and regulated power conversion. (Zhou, Wang, Lukic, Bhattacharya, & Huang, 2009) (Lee, Khaligh, & Emadi, 2009) (Zhou, Lukic, Bhattacharya, & Huang, 2009) (Du et al., 2011) Systems for onboard charging can make conductive or inductive connections. For charging systems that are conductively connected, the charge port inlet and connector will establish direct contact. (Harper, 2013) It is feasible to transfer power wirelessly in inductive charging systems. (Klontz et al., 1994) (Budhia et al., 2011) The authors delved into the inductive charging systems designed for the initial and secondary basic charging stages. Systems for inductive charging might be portable or stationary. (Throngnumchai et al., 2011) (Sergeant & Van den Bossche, 2008) (Boys et al., 2000) Creating off-board charging solutions will allow for higher charging rates. Several studies are technically assessing the effects of EVs, focusing on the power grid, ecology, and economy. The economic impacts of electric vehicles (EVs) may be examined from two perspectives: the perspective of EV owners and the perspective of the utility power grid. To comprehensively assess the economic and environmental effects of electric vehicles (EVs), it is necessary to have a deep grasp of the composition of the energy-generating mix. This also examines the social implications and future of EVs.

In Summary, Section 1 provides a brief introduction to the research report; Section 2 provides a comprehensive review of existing research papers and studies related to Electric Vehicles (EVs); Infrastructure and Charging Levels for Electric Vehicles (EVs) is depicted in section 3; It provides a comprehensive understanding of the infrastructure and charging levels necessary to support the widespread adoption of EVs; Section 4 examines the charging codes and international standards governing Electric Vehicles (EVs); Essential safety and isolation requirements for Electric Vehicle (EV) chargers are outlined in Section 5. Various charging methods and battery charging techniques employed for Electric Vehicles (EVs)are explored in Section 6; In Section 7 evaluated the environmental, social, and economic impacts of Electric Vehicles (EVs); Findings of our research are presented in Section 8. This final section summarizes the key findings and takeaways from the research, highlighting the main conclusions and implications for the future of Electric Vehicles (EVs).

2. LITERATURE SURVEY

H.S. Das et al., Transportation electrification was one of the main research areas over the past decade. Electric vehicles (EVs) gradually took over the market share of conventional internal combustion engine vehicles. The increasing popularity of EVs resulted in a higher number of charging stations, which had significant effects on the electricity grid. Different charging strategies, as well as grid integration methods, were developed to minimize the adverse effects of EV charging and to strengthen the benefits of EV grid integration.

In this paper, a comprehensive review of the EV market, standards, charging infrastructure, and the impact of EV charging on the grid was presented. The paper introduced the current EV status and provided a comprehensive review of important international EV charging and grid interconnection standards. Different infrastructure configurations in terms of control and communication architectures for EV charging were studied and evaluated. The electric power market was analyzed by considering the participation roles of EV aggregators and individual EV owners, and different optimization and game-based algorithms for EV grid integration management were reviewed. The paper particularly presented an evaluation of how future EV developments, such as connected vehicles, autonomous driving, and shared mobility, would affect EV grid integration. It also examined how the development of the power grid moved toward a future energy Internet and how EVs affected and benefited this evolution. Finally, the challenges and suggestions for the future development of EV charging and grid integration infrastructure were evaluated and summarized.

Mohd Rizwan Khalid et al., The large penetration of electric vehicle (EV) charging stations into the existing utility grid brought many power-quality problems, which significantly affected load performances at both large and small consumer ends. This trend grew at a more pronounced rate due to the shift from internal combustion (IC) engine-based vehicles to EVs because of their superiorities. EV chargers converted AC power to DC power for charging EV batteries. During this power conversion by power electronics high frequency switching converters, harmonics were injected into the grid, degrading the power quality of the line current in the grid. Such injection of harmonics caused electrical and thermal overloading of distribution transformers, thereby affecting their life expectancy. Other issues included drawing excess current from the grid, unbalancing demand and supply in the utility grid, and causing voltage deviations in the grid. Consequently, it became essential to maintain the IEEE-519 power quality standards during EV charging. This paper presented a survey of fast charging stations for EVs, their impacts on the existing utility grid, and solutions to overcome these issues. Muhammad Ashfaq et al., Electric vehicles (EVs) had a huge potential to dominate conventional combustion engine vehicles and become the future of road transport while conserving the planet from threatening calamities caused by global warming. EVs offered several benefits over internal combustion engines, particularly in terms of lower greenhouse gas emissions, higher energy efficiency, and reduced dependency on fossil fuels. Despite these advantages, range anxiety and inadequate charging infrastructure were significant limiting factors for the rapid adoption of EVs. This paper presented a thorough literature review on the latest deployment and challenging issues in the implementation of EV infrastructure, charging power levels, and various charging methods. It also examined the impacts of EVs on the electric grid.

Wajahat Khan et al., The idea of electrified transportation was envisioned as a promising approach to alleviate the issue of climate change. However, the cost-effective commercialization and wide acceptance of electrified transportation necessitated the development of an economical, reliable, and fast charging infrastructure. The incorporation of a well-defined electric vehicle (EV) fast charging infrastructure in the smart grid offered numerous potential opportunities, particularly from the perspective of vehicle-to-grid (V2G) technology as well as a solution to the renewable energy intermittency issue. This work reviewed the state-of-the-art technology for fast charging of electric vehicles with a comprehensive coverage of the research work done related to the challenges and roadblocks encountered in its implementation. A comparison of conductive and inductive charging systems was presented, followed by an overview of the charging standards and a discussion on the topologies presented in the literature for fast charging stations. A detailed account was provided concerning the issues and opportunities in the integration of fast charging infrastructure with the grid. The economic aspects of fast charging infrastructure, along with an overview of prospective areas for future research in this field, were also presented.

Salman et al., This paper presented a detailed review of vehicle-to-grid (V2G) technology, in conjunction with various charging strategies of electric vehicles (EVs), and analyzed their impacts on power distribution networks. It was shown in this study that a vehicle, equipped with the ability for a V2G application, offered various features such as regulation of active power, support for reactive power, load balancing, and current harmonics filtering. However, the V2G technology also created challenging issues, such as battery degradation, communication overhead between an EV and a grid, and changes in the entire infrastructure of a distribution network. In particular, the impacts of EVs based on their penetration levels and charging profiles were discussed in detail. Moreover, an extensive analysis was presented on coordinated/uncoordinated charging, delayed charging, off-peak charging, and intelligent scheduling in a distribution network (DN). This study also demonstrated that the economic benefits of V2G technology heavily depended on charging strategies and vehicle aggregation. Table 1 summarizes the key findings and contributions of previous research studies related to Electric Vehicles (EVs),

Table 1. Literature review of previous work

Year	Title	Authors	Methodology	Main Findings	Impact on Infrastructure & Grids
2024	A Survey on Battery Charging Techniques for Electric Vehicles	Zhang et al.	Systematic review of charging methods and trends	Provides an overview of the latest battery charging techniques, from standard to fast charging, and their advancements.	Discusses the effects of high-speed charging on grid stability, with the potential for significant load peaks during peak hours.
2023	Smart Charging Techniques for Electric Vehicles: A Review	Liu & Kumar	Review of smart charging protocols and technologies	Focuses on smart charging systems that optimize energy usage by incorporating real-time data and communication with the grid.	Highlights the potential of reducing grid stress by charging EVs during off-peak hours and using renewable energy sources.
2022	Impact of Fast Charging on Power Grids: Challenges and Solutions	Patel et al.	Simulation modeling of fast charging networks	Analyzes the impact of widespread fast charging stations on grid reliability, highlighting load spikes and the need for advanced infrastructure solutions.	Identifies the need for grid reinforcement, storage solutions, and smart load management to handle the load from fast chargers.
2021	Electric Vehicle Charging Stations: Design, Optimization & Challenges	Zhou et al.	Case study on urban charging station planning	Examines how to optimally place EV charging stations in urban areas and the challenges of providing sufficient power.	Discusses how optimal station placement and load balancing can reduce strain on the grid and improve efficiency.

continued on following page

Table 1. Continued

Year	*Title*	*Authors*	*Methodology*	*Main Findings*	*Impact on Infrastructure & Grids*
2020	*Impact of EV Battery Charging on Local Distribution Networks*	*Smith & Jones*	*Power flow analysis using grid simulation*	*Investigates the local distribution networks under different charging loads and identifies potential issues like voltage drops and overloading.*	*Suggests upgrades to the local distribution network and proposes energy storage solutions for peak shaving.*
2019	*The Role of Vehicle-to-Grid (V2G) in Future Electric Vehicle Charging*	*Huang et al.*	*Analytical study on V2G systems*	*Explores how vehicle-to-grid technology allows for bi-directional energy flow between EVs and the grid, helping with load balancing and energy storage.*	*Emphasizes V2G as a solution to mitigate grid instability, with potential for supporting grid resilience and renewable integration.*

3. INFRASTRUCTURE AND CHARGING LEVELS FOR EVS

This investigation starts by providing a comprehensive examination of the charging infrastructure for electric vehicles (EVs), along with an exploration of different power Modes that are endorsed by different laws. The implementation and progress of electric vehicle supply equipment (EVSE) and charging facilities provide many key challenges that need to be addressed. These issues include the standard nature of charging places, the excess of time and its extent on government policies related to charging demand and distribution, and the need for regulatory control procedures. The presence of charging facilities may result in decreased expenses and requirements for onboard energy storage devices. (Harper, 2013) (Brown et al., 2024). Table 2 lists the EVSE power supply measurements. The important components of an EVSE are listed below:

- Charging connectors and codes for Electric Vehicles are available.
- Charge station locations in public and private spaces
- Various plug points and wiring accessories are required for making the electric link.
- Power outlets and protective equipment

Table 2. EVSE power supply measurements

Electric Vehicle Supply Equipment Types	Electricity source	Power of the charger	Duration of charging	Charging technique
Level 1 charging	120 VAC & 12A to 17A	2.44kW to 2.2kW	15 Hours	Slow charge technique
Level 2 charging	208 to 240VAC & 15A to 90A	4 kW to 20.1 kW	6 hours	Fast charge technique
Level 3 charging	300 to 600VDC and Maximum 400A	120kW up to 240kW	0.5 hour	Fast charging technique

A commonly employed methods for confirming the presence of all essential components are a personalised cable set and a box set placed on an elevated surface. Designs differ across countries, and the location can also impact the design due to important factors such as electric grid connections, voltage, frequency, and transmission system requirements. (Yilmaz & Krein, 2012) (Su et al., 2011) (Nguyen et al., 2024). According to the SAE Standard (Boys et al., 2000), the vehicle must be able to support the EVSE (Electric Vehicle Supply Equipment) for electricity charging levels 1 and 2. Based to the Electric Power Research Institute (EPRI), most Electric Vehicle (EV) owners prefer to charge their cars at home during their sleep hours. However, Figure 4 suggests placing the EVSE outside the vehicle for power level 3. Public stations should utilize power levels 2 and 3 to expedite the charging process at business terminals. It is (Shahed & Rashid, 2024) (English, 2024).

Figure 4. Power levels and charging systems for electric vehicles (EVs).

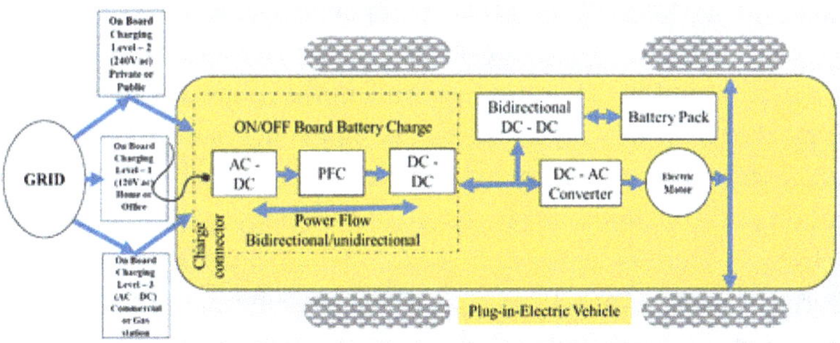

4. CHARGING CODES AND INTERNATIONAL STANDARDS FOR ELECTRIC VEHICLES

The evaluation of electric vehicles (EVs) in the next few years is directly influenced by the adoption of fresh global regulations and charging rules, the provision of sufficient facilities and associated equipment, and the fulfilment of user-friendly software demands at public and private locations. The evaluation of numerous technical issues connected to EVs is now possible thanks to the implementation of international standards and safety requirements. Electric vehicle hardware specifications and infrastructure costs for charging them are connected. (Qadir et al., 2024) (Guzek et al., 2024).

The cost and complexity of the charging infrastructure will rise relative to the current electric infrastructure as a result of future advancements connected to these EV standards and charging rules. According to Article 625-18 of the National Electrical Code, it is mandatory to disconnect and remove power from cables and connectors used for charging levels 2 and 3 before connecting them to an automobile. As a result, EVSE costs will increase. Manufacturers of EVs typically install an interlock to forbid operating the vehicle while it is charging. (Gordon et al., 2017) (Shaukat et al., 2018).

5. REQUIREMENTS FOR ISOLATION AND SAFETY FOR EV CHARGERS

It is necessary to separate all important components of electric vehicles, such as power electronics like DC-to-DC converters, electrical motor driving inverters, high-voltage batteries, and the charging component that connects to the electrical grid. The electrical transformer is the primary component connecting the EVSE to the current grid system. It is necessary to keep the body of the electric vehicle (EV) grounded at all times throughout the procedure of charging, whether using onboard or off-board chargers. Battery separation is essential, and separation (Sethuraman & Rudhramoorthy, 2024) (Can, 2024) is also needed. International standards for Electric Vehicles are shown in Figure 5.

Figure 5. International standards for electric vehicles

SOCIETY FOR AUTOMOBILE ENGINEERS (SAE)	**NATIONAL FIRE PROTECTION ASSOCIATION (NFPA)**
• J1772: EV conductive connector/charging method • J2894: Issues of power quality • J2836/2847/2931: Communication purposes • J1773: Inductive coupled charging • • J2293: For energy transfer systems to find the requirements for EVs and EVSE	• NFPA 70: Safety management • NEC 625/626: Charging systems for EVs • NFPA 70E: For safety • NFPA 70B: Maintenance for electrical equipment
INSTITUTE OF ELECTRICAL AND ELECTRONIC ENGINEERS (IEEE) • IEEE 2030.1.1: Quick DC charging for EVs • IEEE P2690: Charging network management, Vehicle authorization • IEEE P1809: Electric transportation guide • IEEE 1547: Interconnecting electric system with distributed resources/Tie Grid • IEEE 1901: Provide data rate while vehicles charged overnight. • IEEE P2030: Interoperability of smart grid	**INFRASTRUCTURE WORKING COUNCIL (IWC) UNDERWRITERS LABORATORIES INC.** • UL 2231: Safety Purposes • UL 2594/2251,2201: EVSE
INTERNATIONAL ELECTROMECHANICAL COMMISSION (IEC) • IEC-1000-3-6: Issues of power quality • IEC TC 69: Regarding infrastructure of charging and safety requirements • IEC TC 64: Electrical installation, electric shock protection • IEC TC 21: Regarding battery management	**DEUTSCHES INSTITUT FUR NORMUNG** • DIN 43538: Systems for batteries • UNDERWRITERS LABORATORIES INC. • UL 2231: Safety Purposes • UL 2594/2251,2201: EVSE
JAPAN ELECTRIC VEHICLE ASSOCIATION • JEVS C601: EVs charging plugs • JEVS D701: Batteries • JEVS G101-109: Fast Charging	**INTERNATIONAL ORGANIZATION FOR STANDARDIZATION** • ISO 6469-1:2009: Used for on-board rechargeable energy storage systems • ISO/CD 6469-3.3: Safety specifications

In overall, non-isolated DC to DC converters provide notable advantages, including cost reduction, improved efficiency, simplified construction, smaller size and weight, and increased dependability. Unfortunately, the DC/DC converter step lacks galvanic isolation for low-frequency approaches. (Yilmaz & Krein, 2012)

Consequently, a transformer that operates at the same frequency as the power grid is necessary to provide electrical isolation between the grid and the batteries. In order to decrease the number of magnetic materials and the volume requirements, it is necessary to increase the working switching frequency. Off-board locations frequently use battery chargers due to several factors, such as their increased size, the weight increase from the converter cooling system, the need for inductors and capacitors, and the use of a transformer. The design of transformers is essential for decreasing many important attributes, including dimensions, damages, and expense.

For safety reasons, it is suggested to use galvanic isolation instead of on-board isolated chargers while charging circuits, since it is a more cost-effective option. Standards for personnel protection systems cover both isolated and grounded circuits. Table 3 displays a list of the adopted technical regulations and standards that solely focus on EV safety considerations. Table 4 displays the emerging technologies for EV vehicles.

Table 3. List of the scientific guidelines and requirements the security of Electric Vehicles (EVs)

SDO	DOCUMENT	REGION	TITLE
SAE	J2758	North America	Determination of the maximum Available Power from a Rechargeable Energy Storage System on a Hybrid Electrical Vehicle
SAE	J2344	North America	Guidelines for Electrical Vehicle Safety
SAE	J1715	North America	Hybrid Electrical Vehicle (HEV) and Electrical Vehicle (EV) Terminology
UN	ECE324 Regulation 100 Revision 2	International	Uniform provisions concerning the approval of vehicles with regard to specific requirements for the electric power train
SAC	GB/T 18384.2-2015	China	Electrically propelled road vehicles. Safety specifications. Part 2: Vehicle operational safety means and protection against failures
SAC	GB/T 18384.3-2015	China	Hybrid electrical vehicles safety specification
SAC	GB/T 19751-2005	China	Safety requirements on hybrid EV, working voltage up to 600V AC or 1000V DC
SAC	GB/T 4094.2-2005	China	Electrical vehicles. Symbols for controls, indicators and tell-tales
SAC	GB/T 19596-2004	China	Terminology of electrical vehicles
SAC	GB/T 18385-2005	China	Electric vehicles. power performance. Test method
SAC	GB/T 18386-2005	China	Electrical vehicles. Energy consumption and range. Test procedures
SAC	GB/T 18388-2005	China	Electric vehicles. Engineering approval evaluation program
SAC	GB/T 19750-2005	China	Hybrid electric vehicles. Engineering approval evaluation program
AISC	AIS 049	India	Type Approval of Electric Vehicles - Brake performance, grade ability, pass-by-noise level, EMI, Wiper, Lighting system, Safety belt, Steering column, Dashboard etc.
AISC	AIS 102(part 1)	India	Type Approval Hybrid electric vehicles - Category L, M, N (GVW<3500kg)

Table 4. Emerging technology standards

SDO	DOCUMENT	REGION	TITLE
SAE	J3016	North America	Levels of Driving Automation
SAE	J2990	North America	Hybrid EV First and Second Responder Recommended Practice
SAE	J2931/1	North America	Digital Communication for Plug in Electrical Vehicles
SAE	J2931/7	North America	Security for Plug in Electrical Vehicle Communications
SAE	J3061	North America	Best practices for Cyber security
ISO	15118 Series	International	Road vehicles - Vehicle to grid communication interface, including Parts 1,2 and 3

6. THE CHARGING METHODS AND BATTERY CHARGING TECHNIQUES

A battery charger is used to recharge the battery packs in various kinds of electric vehicles, such as plug-in hybrid Electric Vehicles (PHEVs), by supplying them with energy from an external power source. Figure 6 depicts the fundamental configuration of battery chargers.

Figure 6. The basic configuration of a battery charger.

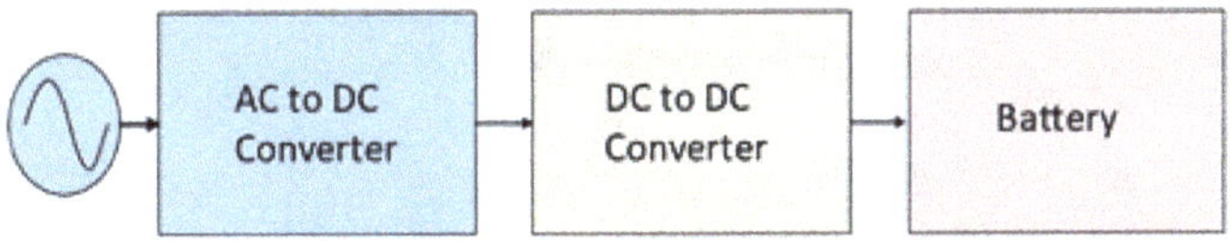

A battery charging device utilises energy to transmit it to the battery while simultaneously monitoring and controlling the electric current to recharge the battery of an electric car. Due to the fact that electric vehicle (EV) batteries function on direct current (DC), a charger is necessary during the whole charging method in order to convert the alternating current (AC) received from the power grid. The configuration of the EV charger includes the necessary DC power level to charge the EV battery. An AC/DC converter is a frequently used architecture for the EV charger. The EV charger now has a converter that converts DC to AC, improving energy conversion and enabling fast charging.

6.1 EV Charging Systems

The technique of energy transfer enables us divide the infrastructure for electric vehicle charging into three distinct groups: first one is conductive electrical charging systems, second one is inductively charging systems final one is battery-swapping circuits. The supply network and the electric vehicle necessitate a tangible linkage for the conductive systems. Power transmission takes place in the conductive mode by directly connecting the EV connection and the charging input port via a cable. This charging technique utilises a cable conductor to make a power connection with electronic devices and boost the generation of energy. It may be used with charging power level outlets, including Level 1, Level 2, and Level 3 charging locations. Electricity-conducting systems are becoming increasingly streamlined and effective. They can serve as onboard chargers for both the gradual charging methods used in Electric Vehicles (EVs) and external chargers for rapid charging procedures. The user's text is " (Mbusi, 2024)". Inductive reasoning charging systems, sometimes referred to as wireless battery charging systems, are an innovative technology that obviates the need of a physical connection to the power grid and the electric vehicle., as seen in Figure 7. An electromagnetic field transmits the electricity to the electric vehicle (EV) battery.

Figure 7. The fundamental configuration of an inductive charging system

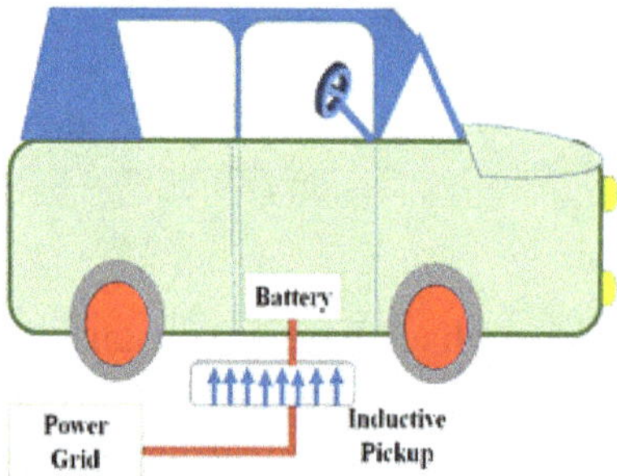

Nevertheless, the converter design on the grid side exhibits higher amounts of harmonic content in its current outputs. The technology for changing batteries forms the foundation of the third major charging system type. Using this method, EV users can swap out their partially or completely exhausted batteries for fully charged ones. At certain sites known as battery switching stations (BSS), the battery switching procedure takes place. TESLA, the dominant player in the transportation industry, reports that the current duration for replacing the battery of an electric vehicle (EV) is a mere 90 seconds. (NGroom, 2013) Battery switching is a method used to decrease or prevent sudden increases in demand for electric vehicles (EVs), which may result in significant cost savings. (Zheng et al., 2013). The existing BMS does not confirm battery safety to the required standards. (Liu et al., 2011) China tops the list of countries with the most BSSs and charging stations worldwide. (Zentani et al., 2024)

6.2 BATTERY CHARGING METHODS FOR EVS

The battery's rechargeable condition is believed to be the main energy source for recharging EVs. One key factor in the current widespread adoption of EVs. At the moment, lithium-ion batteries have replaced lead-acid batteries as the primary energy storage technology. Researchers are conducting several ongoing studies to develop such lithium-sulphide batteries. (Fotouhi et al., 2016). The batteries of electric vehicles (EVs) may be recharged in a variety of ways, including via the use of constant voltage (CV), constant current (CC), continuous power (CP), floating and taper charging, trickle current (TC), and constant voltage (CV). Now, researchers combined the aforementioned methods into a new advanced battery charging technology method, known as the CC-CV charging method, to improve battery management. (Manoharan et al., 2024) (Shahed & Rashid, 2024).

6.2.1 Slow Charging

Constant current (CC) charging is a straightforward charging method that regulates the charging voltage in order to maintain a continuous flow of current towards the battery. Continue the sequence until the battery voltage reaches the desired level. This technique injects a lower current level into the discharged battery. (Shahed & Rashid, 2024). However, overcharging a battery can led to problems with overheating and gassing. A battery pack may suffer serious harm from repeated overcharging or undercharging. It is crucial to regulate battery charging according to the current and voltage parameters. (Manoharan et al., 2024) (Shahed & Rashid, 2024) (English, 2024). Frequently, we denote the present flow of electricity as 10% of the battery's designated maximum capacity. The optimal choices for the CC charging technique are nickel-cadmium and nickel-metal hydrate batteries. The CV charging method prevents excessive power consumption. Furthermore, the CV charging method controls the threshold of battery charging when a state of charge by using less current throughout the charging process. (Manoharan et al., 2024) (Shahed & Rashid, 2024) (English, 2024).

The CC charging technique significantly increases the charging time to avoid problems. When using a steady, unregulated voltage source to taper charge, the charging current reduces uncontrollably because of the high cell voltage caused by the faster charging rate. Overcharging could severely damage the battery pack as a result. In contrast, the oat charging methodology uses the CV charging method under the battery's top border. Typically, backup emergency power situations use this technique, which works with lead-acid batteries. Trickle current charging overcomes battery discharge by charging the EV battery with a lower current.

6.2.2 Fast Charging

Kang et al. (2013) have developed various charging techniques to achieve faster and safer charging of EV batteries and to address the aforementioned issues. Combine the three main segments of constant current (CC) and constant voltage (CV) charging techniques: the first segment begins with a pre-charging mode, the second employs CC charging, and the third uses CV charging. In the pre-charge mode, we gradually increase the current to raise the battery voltage to a certain threshold. We do this to ensure a regulated introduction of power throughout the initial process, thereby preventing battery damage. Controlling the charger produces a higher value of constant current than the threshold limit. As a result, the battery rapidly charges and attains its State of Charge (SOC) of 80%. (Manoharan et al., 2024) (Shahed & Rashid, 2024) (English, 2024) (Kang et al., 2013).

The battery charger now enters the CV charging mode to control current flow and safeguard the battery from overcharging damage. The improved energy density and power of Li-ion batteries enable rapid charging via the use of the CC-CV battery charging technology. Most chargers available on the market use the CC-CV charging mechanism. The benefits of using the CC-CV charging approach are as follows: The key factors for efficient battery charging include: 1) Restricting the voltage during charging; 2) Restricting the current during charging; 3) Regulating the power pumped into the battery; 4) Properly using the battery controller; 5) Safeguarding the battery against excessive voltages; 6) Minimising thermal stress; and 7) Achieving fast charging. However, the addition of rest stages increases the EV battery's total charging time. (Al-karakchi et al., 2015) When deploying electric vehicles (EVs) on a wide scale, the duration required to charge the battery becomes a crucial factor for the battery management system. Multiple research studies have shown that the core CC-CV charging technique is more efficient in terms of both batteries charging capacity and total charging time for electric vehicle (EV) batteries. Therefore, several technical research and development projects employ the PC charging method. The battery's current charging setup still has a lot of technical issues, though.

6.3 Charging Technique Based On Vehicle Mobility

These vehicles come equipped with both dynamic and static functional charging methods. The static and wireless charging (SWC) technology may charge an electric vehicle (EV) even while it is in motion. The state of SWC technology is gradually improving. Figure 8 illustrates the roadway model with various charging techniques.

Figure 8. Roadway model of different types of charging techniques

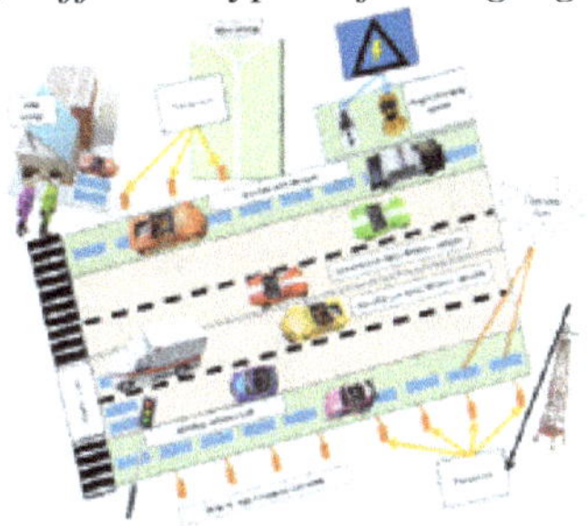

6.3.1 Static Charging (SC)

6.3.1.a. Static plug-in charging (SPC)

Electric Vehicles (EVs) may be charged using four distinct methods: Mode 1, Mode 2, Mode 3, and Mode 4. We create categories based on the battery's performance and charging time. It is believed that mode 1 charging is incredibly slow and takes the entire night. For mode 2 charging, the words nominal and semi-fast are commonly utilized. Single-phase electricity facilitates charging in modes 1 and 2. Households commonly implement mode 1 charging, while public areas such as malls commonly use mode 2 charging. We refer to modes 3 and 4 as rapid charging because they can fully charge a battery in less than 30 minutes and often require a three-phase supply. Charging technologies fall into three main categories: fast/ultra-fast charging, domestic charging, and public charging. However, Figure 9 describes the special features and limitations associated with each of these charging methods.

Figure 9. The features of charging technologies.

Mode 1	Mode 2	Modes 3 & 4

Slow Charging	Medium Fast Charging	Ultra-Fast Charging
Home charging	Parking Places and shopping malls	Public Charging Station
7-15 Hours	3-6 Hours	20-40 Minutes
63A Max Current	32A Max Current	16A Max Current

6.3.1.b Static Wireless Charging (SWC)

However, defining precise requirements for static WPT systems used to charge light-duty Electric Vehicles has only recently become possible. Following over twenty years of increasing research interest in electric vehicle (EV) wireless charging systems, the automotive industry is becoming more and more interested in wireless power transfer (WPT), a method of recharging employing resonant-coupled inductors. Each of these solutions has unique properties, such as cost-effectiveness, a decrease in a stray magnetic field, and resilience to misalignment. Numerous researchers have concurrently offered numerous coil topologies and forms, which sets them apart.

6.3.1.c. Working Principle of A WPT Systems

SAE RP proposes a generic concept for a WPT system. In Figure 10, its primary power sources are displayed. On the base (or transmitter) portion of the entire system, a DC/AC converter, often an H-bridge, powers the transmitting coil and adjusts the parts. In addition, this converter serves as the power factor corrector (PFC) for the system. On the vehicle's (or receiver's) side, the receiver coil and its adjusting components connect to a rectifier with a diode and the vehicle's battery. This study does not consider the possibility of an alternative DC/DC converter managing the battery connection.

Figure 10. Schematic representation of the SAE 12954 RP-described power components of WPT systems.

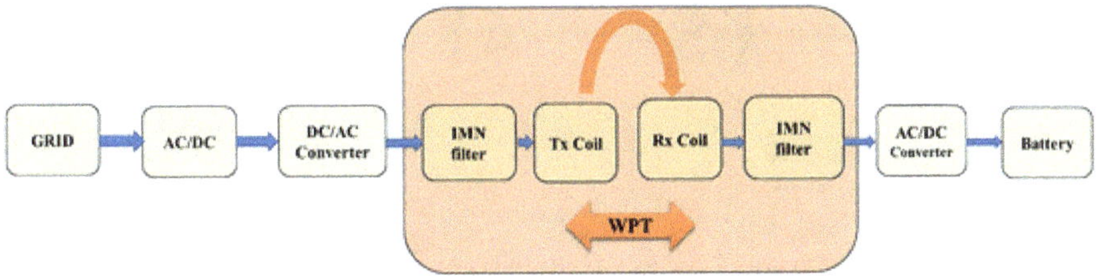

The two coils, along with several ancillary electromagnetic and mechanical components, comprise the magnetic component that is directly in charge of wireless power transfer. "Vehicle Assembly" (VA) and "Ground Assembly" (GA) are the terms used to describe these groups. We refer to the blocks housing the compensation parts as impedance matching networks (IMN) because they can be as basic as a single capacitor circuit or as complex as a circuit with several reactive elements. Currently, it is believed that each IMN block consists of just one adjusting capacitor connected in line with each coil. This assumption simplifies the analysis and provides a commonly used topology.

6.3.2 Dynamic Wireless Charging (DWC)

Recent developments in Dynamic Wireless Power Transfer (DWPT) technology, as seen in Figure 11, have made it possible to extend the distance that Electric Vehicles (EVs) can travel. The technology of on-road wireless charging decreases the size of the electric vehicle (EV) battery and charges the automobile while it is moving. As a result, electrified roadways could lessen the need for large batteries in electric vehicle applications. We must engineer the entire DWPT system efficiently to ensure maximum power transmission from ground level to the vehicle.

Figure 11. A designed Model DWC system

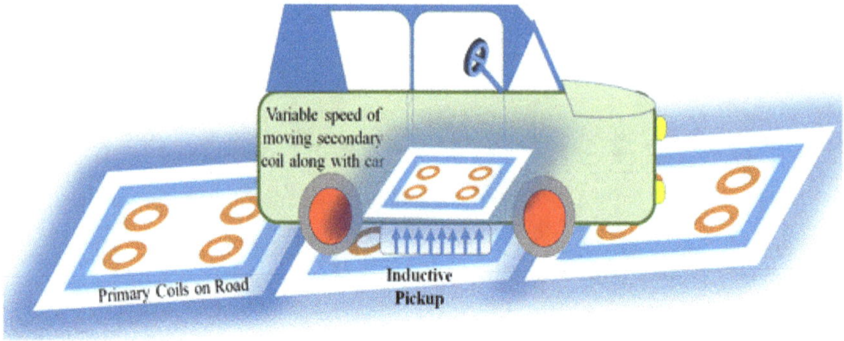

6.3.3 Dynamic Combined with Static Wireless Charging (QDWC) Method

These vehicles have both dynamic and static functional charging methods fixed. The SWC method allows for charging an EV while it is in motion. The state of SWC technology is gradually improving.

7. IMPACT ASSESSMENT OF EVS

Given the current state of the transportation sector, projections indicate that EVs' rapid development rate will soon significantly penetrate distribution networks. The existing power networks may face increased demands due to the significant charging requirements of Electric Vehicles (EVs), which are having a detrimental effect on the current traditional distribution grids. Nevertheless, the extensive utilisation of contemporary Electric Vehicles (EVs) yields favourable outcomes for both the economy and the environment. By connecting them to a power grid, we can access several interesting capabilities of Electric Vehicles (EVs). The mentioned features include load distribution, provision of reactive electricity, control of active power, and facilitation of renewable energy sources. Additionally, EVs can reduce emissions of greenhouse gases. (Al-Sahlawi et al., 2024) (Jain & Bhullar, 2024).

The potential benefits of switching to electric cars include cheaper import costs for oil, more jobs, market expansion, new enterprises, fresh air, and less greenhouse gas (GHG) emissions is shown in Figure 12. Less employment in specialized industries and a decline in tax revenue from the use of fossil fuels would result in a decline in the value of some well-known companies. Every significant technological advancement comes with benefits and drawbacks, some of which are transient and others of which are permanent. (Jain & Bhullar, 2024) (Chigbu, 2024)

Figure 12. EV impact types

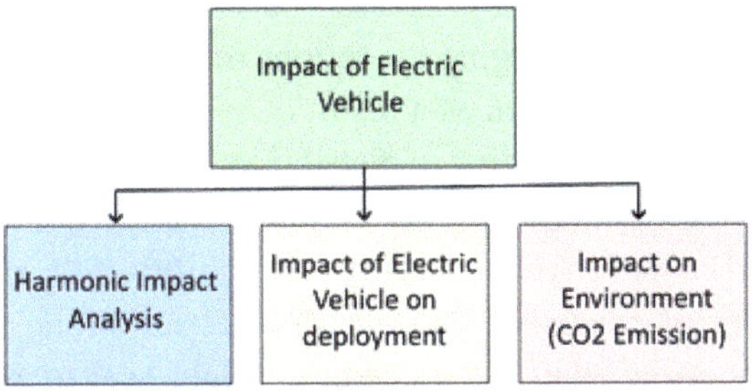

7.1 Impacts of Electric Vehicles on Deployment

To address serious air pollution challenges and lower carbon dioxide (CO_2) emissions, recently formed EV businesses are promoting the use of electric cars (EVs) and decarbonising the power sector. (Aktar et al., 2024)

The utilisation of Electric Vehicles (EVs) has two primary impacts on load management strategies: (a) either the use of EVs or load management strategies reduces the effectiveness of load factor improvement due to their interaction, or (b) the use of EVs unintentionally increases peak demand. (Mohammed et al., 2024)

Several research studies have examined the effect of electric vehicle (EV) charging techniques on the load profile and peak demand of a distribution system. Diverse distribution networks across many countries conduct these impact evaluations. We did an investigation to analyse a significant rise in demand caused by the widespread adoption of Electric Vehicles (EVs). The authors discovered that when they use uncoordinated charging techniques and adopt 100% Electric Vehicles (EVs), the peak demands on days with normal loads surpass the power generation capacity. The distribution network must include additional production capacity to accommodate up to 93% of the electric vehicle demand during off-peak periods (Kyriakou et al., 2024). The peak demand experienced a 53% rise due to the lack of coordination in charging, given that Electric Vehicles (EVs) accounted for 30% of the total vehicles. (Kyriakou et al., 2024) The likelihood of choosing the peak load hours rises due to the erratic charging habits of EV owners. As a result, the widespread adoption of EVs under various charging conditions will significantly increase peak demand for electricity. (Ali et al., 2020) (McGarry et al., 2024) (Chakraborty & Pal, 2024) (Yuvaraj et al., 2024). Thankfully, the power network's peak demand and load profile problems may be addressed via the use of optimal charging solutions, time-of-use schedules, and integration of renewable energy sources (ESCAP, 2024).

7.2 Analysis of the Impact of Harmonics

The electric vehicle (EV) charging operations might lead to power quality problems in the network of distribution as a result of switching phenomena. Harmonic distortion greatly affects the de-rating of distribution components; hence it is crucial to take this into account. The widespread adoption of electric vehicles (EVs) might have a significant influence on the reliability and stability of the power grid's supply. The variable number of electric vehicle (EV) batteries, along with the uncertain energy use throughout the charging procedure, might lead to a challenge in managing the demand on the power grid. Equations (1) and (2) enable the calculation of the total harmonic distortion (THD) values for current and voltage, expressed as percentages. (Nutkani et al., 2024)

$$THD_i = \frac{\sqrt{\sum_{h=2}^{M} I_{hr}^2}}{I_f} \times 100\% \tag{1}$$

$$THD_v = \frac{\sqrt{\sum_{h=2}^{M} V_{hr}^2}}{V_f} \times 100\% \tag{2}$$

Where hr - the order number of harmonics, M - the maximum number of harmonics occurring, and RMS values of current and voltage at h^{th} are harmonic components.

I_f, V_f - frequency values of current and voltage.

Different charging scenarios for EVs do not reveal any significant harmonic problems. The voltage THD value has increased slightly (less than 1%). (Khalid et al., 2024). We conduct a thorough harmonic analysis using the Monte Carlo simulation approach, taking into account dynamic EV factors such as charging time and location. The findings indicated that the harmonic influence on the distribution network was almost negligible. (Srivastava et al., 2023)

7.2.1 Transformation of the Battery's State of Health

When it comes to charging business cars, it is only possible to do so at their commercial location within certain time intervals, such as overnight or after shifts finish. Because of this, commercial cars charging on their own property must be able to fulfil their daily transportation needs without triggering anxiety about range. Therefore, the charging and discharging processes of commercial vehicles reduce the average amount of driving that estimated timings (ETs) can accomplish, potentially impairing battery health and limiting its lifespan and optimal performance.

7.3 Impacts of Electric Vehicles on the environment (CO_2 emissions)

The effect of transportation on the environment is one of the world's major problems, having an impact on both the environment and human life, both directly and indirectly. The transportation sector, on the other hand, is one of the biggest energy consumers. Environmental problems increase as a result of CO_2 production, which is what drives global warming. (Tintelecan et al., 2019). We used experimental data from the test vehicle and a test battery cell to study the impact of battery deterioration on CO_2 emissions during operation. As the deterioration progresses, the batteries' internal resistance rises. Internal resistance can have an impact on a battery's operation. The internal resistance will cause a shift in energy consumption from its idle state. As a result, CO_2 emissions will rise. Batteries that have degraded lose capacity. Reduced battery capacity may shorten the range of electric cars. As a result, cars become less useful, which may serve as motivation for battery replacement. We conducted a study to examine the effect of battery deterioration on energy consumption and range by measuring the real fluctuations of a test electric car's energy use. According to these viewpoints, battery deterioration could increase CO_2 emissions while in use. (Takahashi et al., 2023) (Rietmann et al., 2020) (Pan et al., 2023).

8. RESULTS AND DISCUSSIONS

In this part, we report the real-time findings from the RT-Lab simulator for the proposed three-tiered billing systems. The entire system's battery is 10,000 Wh powerful and 40 Ah in length. The simulation provides a detailed representation of several parameters including the battery state of charge (SOC), reactive power, DC bus voltage, PFC inductor current, power, and phase shift. Figure 13 illustrates Level 1 and 2 charging the charging process starts with a steady current of 6 A and gradually increases voltage. As soon as the voltage hits the 262 V threshold, the current will begin to drop. Level 1 charging uses a 310 V DC bus voltage. This study proves that the CC/CV charging technique is effective. Level 2 charging, as shown in Figure 13, likewise consists of two stages: CC and CV. Charging enters CV mode at 272 V and 310 V on the DC bus, with a current of 41 A at the CC stage. As the charging power drops in CV mode, the inductor and phase shift both fall in tandem. Both the voltage and current are constant at 273 V and 82 A, respectively. Reactive power remains at 30 kVAR, and the DC voltage is 360 V. The same period, Figure 13 displays the full charge comparison of the three-level charging methodology. At Level 3, the quickest way to charge is to raise the SOC from 10% to 84%. The SOC surges from 10% to 49% in the level 2 charging. The slowest charger, a Level 1 charger, takes 25 minutes to charge the battery to 5% capacity. Table 5 summarises the numerical findings based on the simulated results. All of the power evaluations, charging currents, and voltages are by SAE standards. A battery's anticipated full charging time is the amount of time it takes to fully charge the battery from zero SOC to one hundred percent. We use the Opmonitor block to capture the model's operations in RT-Lab.

Figure 13. The charging framework for Levels 1 and 2

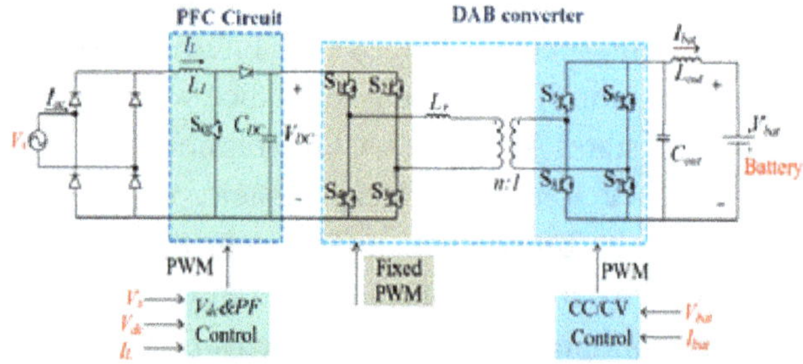

Table 5. Simulation analysis

	Level 1 charging	Level 2 charging	Level 3 charging
Initial SOC in three levels	10%	10%	10%
The initial phase shift occurs	13°	19°	34°
DC voltage	310 V	310 V	360 V
Charging power	1320 W	10950 W	22380 W
Current flow	7 A	42 A	82 A
Voltage flow	265 *V*	275 *V*	280 *V*
End SOC	14.5%	48.6%	82.5%
fully charging time	11 hours	1.2 hour	30 mins

9. CONCLUSION

The review report analyses the global standards for charging methods, power quality problems, and integration with electrical grids, safety limitations, network communications, and equipment maintenance for Electric Vehicles (EVs). The methods of charging we have extensively examined several charging processes for Electric Vehicles (EVs), also with inductive charging, conductive charging, and swapping battery networks. We also discuss quick and progressive battery charging techniques for vehicles that are mobile. This study examines the current state, latest advancements, and most challenging issues in implementing Electric Vehicle (EV) charging and structural systems, adhering to various global standards and charging regulations for Levels 1, 2, and 3 using the RT-Lab simulator, as well as the modes of vehicles charging 1, 2, 3, and 4. Furthermore, it examines the societal implications and future possibilities of electric cars (EVs). We methodically classify and extensively analyse the positive and negative consequences of future technology in transportation, along with strategies to mitigate the adverse effects and maximise the long-term benefits of the positive ones. Vehicle-to-grid technology is mostly based on bidirectional charging. Future technological advancements in EV battery application, charging station design, and safety conditions are accelerating the shift to wireless EV charging in the modern world. This is due to factors such as waiting to charge the battery, reducing battery size, reducing vehicle cost, and considering electricity demand reductions by using renewable resources. The implementation of EVs in transportation has both beneficial and then detrimental effects on the economy and the environment. Nevertheless, it is important to take into account the possible disadvantages and improve charging methods in order to alleviate these issues in a wireless charging setting.

REFERENCES

Aktar, A. K., Taşcıkaraoğlu, A., Erdinç, O., & Güner, S. (2024). Impacts of Distribution-Level Joint Scheduling of Electric Vehicle Battery Charging and Swapping Stations on Reliability Improvement. IEEE Transactions on Industry Applications. https://ieeexplore.ieee.org/abstract/document/10561522

Al-karakchi, A. A. A., Lacey, G., & Putrus, G. (2015, September). A method of electric vehicle charging to improve battery life. In 2015 50th International Universities Power Engineering Conference (UPEC) (pp. 1-3). IEEE. https://ieeexplore.ieee.org/abstract/document/7339846

Al-Sahlawi, A. A. K., Ayob, S. M., Tan, C. W., Ridha, H. M., & Hachim, D. M. (2024). Optimal Design of Grid-Connected Hybrid Renewable Energy System Considering Electric Vehicle Station Using Improved Multi-Objective Optimization: Techno-Economic Perspectives. *Sustainability (Basel), 16*(6), 2491. https://www.mdpi.com/2071-1050/16/6/2491. DOI: 10.3390/su16062491

Ali, H., Hussain, S., Khan, H. A., Arshad, N., & Khan, I. A. (2020, September). Economic and environmental impact of vehicle-to-grid (V2G) integration in an intermittent utility grid. In 2020 2nd International Conference on Smart Power & Internet Energy Systems (SPIES) (pp. 345-349). IEEE. https://www.mdpi.com/2071-1050/14/21/13856

Anegawa, T. (2010). Development of quick charging system for electric vehicle. https://www.osti.gov/etdeweb/biblio/21423136

Bayram, I. S., & Tajer, A. 2017. Plug-in electric vehicle grid integration. Artech House. https://pureportal.strath.ac.uk/en/publications/plug-in-electric-vehicle-grid-integration

Botsford, C., & Szczepanek, A. (2009, May). Fast charging vs. slow charging: Pros and cons for the new age of Electric Vehicles. In *International Battery Hybrid Fuel Cell Electric Vehicle Symposium* (pp. 1–9). Citeseer., https://static1.squarespace.com/static/55f4e4d1e4b029b54a77cf2d/t/55fdf7a9e4b016bd35401ea1/1442707369493/EVS24_3960315_Botsford_r.pdf

Boys, J. T., Covic, G. A., & Green, A. W. (2000). Stability and control of inductively coupled power transfer systems. *IEE Proceedings. Electric Power Applications, 147*(1), 37–43. https://digital-library.theiet.org/content/journals/10.1049/ip-epa_20000017. DOI: 10.1049/ip-epa:20000017

Brown, A., Cappellucci, J., Heinrich, A., & Cost, E. (2024). Electric vehicle charging infrastructure trends from the alternative fueling station locator: Third quarter 2023 (No. NREL/TP-5400-88223). National Renewable Energy Laboratory (NREL), Golden, CO (United States). https://www.osti.gov/biblio/2293493

Budhia, M., Covic, G. A., Boys, J. T., & Huang, C. Y. (2011, September). Development and evaluation of single sided flux couplers for contactless electric vehicle charging. In 2011 IEEE Energy Conversion Congress and Exposition (pp. 614-621). IEEE. https://ieeexplore.ieee.org/abstract/document/6063826

Can, E. (2024). DC–DC converter with multiple inputs and full isolated multi ports charging battery in photovoltaic energy systems. *Sadhana*, *49*(2), 148. https://link.springer.com/article/10.1007/s12046-024-02506-y. DOI: 10.1007/s12046-024-02506-y

Chakraborty, P., & Pal, M. (2024). Planning of fast charging infrastructure for Electric Vehicles in a distribution system and prediction of dynamic price. *International Journal of Electrical Power & Energy Systems*, *155*, 109502. https://www.sciencedirect.com/science/article/pii/S0142061523005598. DOI: 10.1016/j.ijepes.2023.109502

Chandrakar, N., Sekhar, C. C., Mageshwari, S., & Daniel, S. A. (2024). Electric vehicle progression in the society and their consequences. In *Advanced Technologies in Electric Vehicles* (pp. 127–160). Academic Press., https://www.sciencedirect.com/science/article/abs/pii/B9780443189999000053 DOI: 10.1016/B978-0-443-18999-9.00005-3

Chigbu, B. I. (2024). Advancing sustainable development through circular economy and skill development in EV lithium-ion battery recycling: A comprehensive review. *Frontiers in Sustainability*, *5*, 1409498. https://www.frontiersin.org/journals/sustainability/articles/10.3389/frsus.2024.1409498/full. DOI: 10.3389/frsus.2024.1409498

Du, Y., Lukic, S., Jacobson, B., & Huang, A. (2011, September). *Review of high power isolated bi-directional DC-DC converters for PHEV/EV DC charging infrastructure. In 2011 IEEE Energy Conversion Congress and Exposition*. IEEE., https://ieeexplore.ieee.org/abstract/document/6063818

English, R. (2024). Raising Public Awareness of Electric Vehicle Technology to Increase Sales. https://scholarworks.waldenu.edu/dissertations/15388/

ESCAP. U. (2024). Accelerating electric mobility in Public Transport: how Asia and the Pacific could unlock financing for electric buses. https://repository.unescap.org/handle/20.500.12870/6756

Fasugba, M. A., & Krein, P. T. (2011, September). Gaining vehicle-to-grid benefits with unidirectional electric and plug-in hybrid vehicle chargers. In 2011 IEEE vehicle power and propulsion conference (pp. 1-6). IEEE. https://ieeexplore.ieee.org/abstract/document/6043207

Fotouhi, A., Auger, D. J., Propp, K., Longo, S., & Wild, M. (2016). A review on electric vehicle battery modelling: From Lithium-ion toward Lithium–Sulphur. *Renewable & Sustainable Energy Reviews*, *56*, 1008–1021. https://www.sciencedirect.com/science/article/abs/pii/S1364032115013921. DOI: 10.1016/j.rser.2015.12.009

Gordon, L. B., Carr, K. D., & Graham, N. (2017). Complete electrical arc hazard classification system and its application. *IEEE Transactions on Industry Applications*, *53*(5), 5078–5087. https://ieeexplore.ieee.org/abstract/document/7878523/. DOI: 10.1109/TIA.2017.2682158

Guzek, M., Jackowski, J., Jurecki, R. S., Szumska, E. M., Zdanowicz, P., & Żmuda, M. (2024). Electric Vehicles—An Overview of Current Issues—Part 2—Infrastructure and Road Safety. *Energies*, *17*(2), 495. https://www.mdpi.com/1996-1073/17/2/495. DOI: 10.3390/en17020495

Hardman, S., Shiu, E., & Steinberger-Wilckens, R. (2016). Comparing high-end and low-end early adopters of battery Electric Vehicles. *Transportation Research Part A, Policy and Practice*, *88*, 40–57. https://www.sciencedirect.com/science/article/pii/S0965856416302208. DOI: 10.1016/j.tra.2016.03.010

Harper, J. D. (2013). Development and implementation of sae dc charging digital communication for plug-in electric vehicle dc charging (No. 2013-01-1188). SAE Technical Paper. https://www.sae.org/publications/technical-papers/content/2013-01-1188/

IEA. *Energy Policies of IEA Countries: Greece 2011 Review*. OECD Pub., 2011. https://www.iea.org/reports/energy-policies-of-iea-countries-greece-2011-review

International Energy Agency. Global EV Outlook 2018—Towards Cross-modal Electrification; International Energy Agency: Paris, France. https://www.iea.org/reports/global-ev-outlook-2018

Jain, F., & Bhullar, S. (2024). Operating Modes of Grid Integrated PV-Solar Based Electric Vehicle Charging System-A Comprehensive Review. e-Prime-Advances in Electrical Engineering, Electronics and Energy, 100519. https://www.sciencedirect.com/science/article/pii/S2772671124001013

Jin, X., & Meintz, A. (2015). Challenges and Opportunities for Transactive Control of Electric Vehicle Supply Equipment. A Reference Guide. https://www.osti.gov/biblio/1215194

Kang, T., Chae, B., & Suh, Y. (2013, June). Control algorithm of bi-directional power flow rapid charging system for electric vehicle using Li-Ion polymer battery. In 2013 IEEE ECCE Asia Downunder (pp. 499-505). IEEE. https://ieeexplore.ieee.org/abstract/document/6579143

Khalid, M., Thakur, J., Bhagavathy, S. M., & Topel, M. (2024). Impact of public and residential smart EV charging on distribution power grid equipped with storage. *Sustainable Cities and Society*, *104*, 105272. https://www.sciencedirect.com/science/article/pii/S221067072400101X. DOI: 10.1016/j.scs.2024.105272

Klontz, K. W., Divan, D. M., Novotny, D. W., & Lorenz, R. D. (1994). *U.S. Patent No. 5,341,083*. U.S. Patent and Trademark Office., https://patents.google.com/patent/US5341083A/

Kumar, M., Panda, K. P., Rosas-Caro, J. C., Valderrabano-Gonzalez, A., & Panda, G. (2023). Comprehensive review of conventional and emerging maximum power point tracking algorithms for uniformly and partially shaded solar photovoltaic systems. *IEEE Access : Practical Innovations, Open Solutions*, *11*, 31778–31812. https://ieeexplore.ieee.org/abstract/document/10083133. DOI: 10.1109/ACCESS.2023.3262502

Kyle, P., & Kim, S. H. (2011). Long-term implications of alternative light-duty vehicle technologies for global greenhouse gas emissions and primary energy demands. *Energy Policy*, *39*(5), 3012–3024. https://www.sciencedirect.com/science/article/abs/pii/S0301421511001960. DOI: 10.1016/j.enpol.2011.03.016

Kyriakou, D. G., Kanellos, F. D., & Ipsakis, D. (2024). Multi-agent-based real-time operation of microgrids employing plug-in Electric Vehicles and building prosumers. Sustainable Energy, Grids and Networks, 37, 101229. https://www.sciencedirect.com/science/article/pii/S2352467723002370

Lee, Y. J., Khaligh, A., & Emadi, A. (2009). Advanced integrated bidirectional AC/DC and DC/DC converter for plug-in hybrid Electric Vehicles. *IEEE Transactions on Vehicular Technology*, *58*(8), 3970–3980. https://ieeexplore.ieee.org/abstract/document/5170004. DOI: 10.1109/TVT.2009.2028070

Liu, Y., Hui, F., Xu, R., Chen, T., Xu, X., & Li, J. (2011). Investigation on the Construction Mode of the Charging Station and Battery-Exchange Station. Asia-Pacific Power and Energy Engineering Conference, APPEEC. 1-2. . https://ieeexplore.ieee.org/abstract/document/5777909DOI: 10.1109/APPEEC.2011.5748626

Manoharan, S., Mahalakshmi, B., Ananthi, K., & Elakya, A. (2024, March). A Review on Smartphone Charger: Technologies and Challenges. In 2024 5th International Conference on Intelligent Communication Technologies and Virtual Mobile Networks (ICICV) (pp. 84-88). IEEE. https://ieeexplore.ieee.org/abstract/document/10511183

Mbusi, A. J. (2024). Reimagining Electric Vehicles for a Sustainable Future. https://www.preprints.org/manuscript/202406.1200/v1

McGarry, C., Dixon, J., Flower, J., Bukhsh, W., Brand, C., Bell, K., & Galloway, S. (2024). Electrified heat and transport: Energy demand futures, their impacts on power networks and what it means for system flexibility. *Applied Energy*, *360*, 122836. https://www.sciencedirect.com/science/article/pii/S0306261924002198. DOI: 10.1016/j.apenergy.2024.122836

Mohammed, A., Saif, O., Abo-Adma, M., Fahmy, A., & Elazab, R. (2024). Strategies and sustainability in fast charging station deployment for Electric Vehicles. *Scientific Reports*, *14*(1), 283. https://www.nature.com/articles/s41598-023-50825-7. DOI: 10.1038/s41598-023-50825-7 PMID: 38168937

NGroom. (2013) Electric Car Maker Tesla Unveils 90s Battery Pack Swap. https://www.reuters.com/ article/2013/06/21/ us-tesla-swap-idUSBRE95K07H20130621

Nguyen, D. M., Kishk, M. A., & Alouini, M. S. (2024). Dynamic charging as a complementary approach in modern EV charging infrastructure. *Scientific Reports*, *14*(1), 5785. https://www.nature.com/articles/s41598-024-55863-3. DOI: 10.1038/s41598-024-55863-3 PMID: 38461152

Nutkani, I., Toole, H., Fernando, N., & Andrew, L. P. C. (2024). Impact of EV charging on electrical distribution network and mitigating solutions–A review. IET Smart Grid. https://ietresearch.onlinelibrary.wiley.com/doi/full/10.1049/stg2.12156

Pan, S., Yu, W., Fulton, L. M., Jung, J., Choi, Y., & Gao, H. O. (2023). Impacts of the large-scale use of passenger Electric Vehicles on public health in 30 US. metropolitan areas. *Renewable & Sustainable Energy Reviews*, *173*, 113100. https://www.sciencedirect.com/science/article/abs/pii/S1364032122009819. DOI: 10.1016/j.rser.2022.113100

Pelegov, D. V., & Chanaron, J. (2022). Electric Car Market Analysis Using Open Data: Sales, Volatility Assessment, and Forecasting. Sustainability. https://www.mdpi.com/2071-1050/15/1/399

Qadir, S. A., Ahmad, F., Al-Wahedi, A. M. A., Iqbal, A., & Ali, A. (2024). Navigating the complex realities of electric vehicle adoption: A comprehensive study of government strategies, policies, and incentives. Energy Strategy Reviews, 53, 101379. https://www.sciencedirect.com/science/article/pii/S2211467X24000865

Rawson, M., & Kateley, S. (1999). Electric vehicle charging equipment design and health and safety codes. *SAE Transactions*, •••, 3256–3262. https://www.jstor.org/stable/44733996. DOI: 10.4271/1999-01-2941

Rietmann, N., Hügler, B., & Lieven, T. (2020). Forecasting the trajectory of electric vehicle sales and the consequences for worldwide CO2 emissions. *Journal of Cleaner Production, 261*, 121038. https://www.sciencedirect.com/science/article/abs/pii/S0959652620310854. DOI: 10.1016/j.jclepro.2020.121038

Sergeant, P., & Van den Bossche, A. (2008). Inductive coupler for contactless power transmission. *IET Electric Power Applications, 2*(1), 1–7. https://digital-library.theiet.org/content/journals/10.1049/iet-epa_20070059. DOI: 10.1049/iet-epa:20070059

Sethuraman, R., & Rudhramoorthy, M. (2024). Performance of Bidirectional ON-Board Charger in Electric Vehicle: A Review. e-Prime-Advances in Electrical Engineering, Electronics and Energy, 100613. https://www.sciencedirect.com/science/article/pii/S2772671124001931

Shahed, M. T., & Rashid, A. H. U. (2024). Battery charging technologies and standards for Electric Vehicles: A state-of-the-art review, challenges, and future research prospects. *Energy Reports, 11*, 5978–5998. https://www.sciencedirect.com/science/article/pii/S2352484724003408. DOI: 10.1016/j.egyr.2024.05.062

Shaukat, N., Khan, B., Ali, S. M., Mehmood, C. A., Khan, J., Farid, U., Majid, M., Anwar, S. M., Jawad, M., & Ullah, Z. (2018). A survey on electric vehicle transportation within smart grid system. *Renewable & Sustainable Energy Reviews, 81*, 1329–1349. https://www.sciencedirect.com/science/article/abs/pii/S1364032117307190. DOI: 10.1016/j.rser.2017.05.092

Srivastava, A., Manas, M., & Dubey, R. K. (2023). Electric vehicle integration's impacts on power quality in distribution network and associated mitigation measures: A review. *Journal of Engineering and Applied Sciences (Asian Research Publishing Network), 70*(1), 32. https://link.springer.com/article/10.1186/s44147-023-00193-w

Su, W., Eichi, H., Zeng, W., & Chow, M. Y. (2011). A survey on the electrification of transportation in a smart grid environment. *IEEE Transactions on Industrial Informatics, 8*(1), 1–10. https://ieeexplore.ieee.org/abstract/document/6051485. DOI: 10.1109/TII.2011.2172454

Sun, X., Li, Z., Wang, X., & Li, C. (2020). Technology Development of Electric Vehicles: A Review. *Energies, 90*. Advance online publication. https://www.mdpi.com/1996-1073/13/1/90. DOI: 10.3390/en13010090

Takahashi, R., Negishi, K., Noda, H., & Mizutani, M. (2023). Estimating the Dominant Life Phase Concerning the Effects of Battery Degradation on CO2 Emissions by Repetitive Cycle Applications: Case Study of an Industrial Battery System Installed in an Electric Bus. *Energies*, *16*(3), 1508. https://www.mdpi.com/1996-1073/16/3/1508. DOI: 10.3390/en16031508

Throngnumchai, K., Kai, T., & Minagawa, Y. (2011, September). A study on receiver circuit topology of a cordless battery charger for Electric Vehicles. In 2011 IEEE Energy Conversion Congress and Exposition (pp. 843-850). IEEE. https://ieeexplore.ieee.org/abstract/document/6063858

Tintelecan, A., Dobra, A. C., & Marţiş, C. (2019, October). LCA indicators in Electric Vehicles environmental impact assessment. In 2019 Electric Vehicles International Conference (EV) (pp. 1-5). IEEE. https://ieeexplore.ieee.org/abstract/document/8892893

Van de Graaf, T., & Lesage, D. (2009). The International Energy Agency after 35 years: Reform needs and institutional adaptability. *The Review of International Organizations*, *4*(3), 293–317. DOI: 10.1007/s11558-009-9063-8

Yilmaz, M., & Krein, P. T. (2012). Review of battery charger topologies, charging power levels, and infrastructure for plug-in electric and hybrid vehicles. *IEEE Transactions on Power Electronics*, *28*(5), 2151–2169. https://ieeexplore.ieee.org/abstract/document/6280677. DOI: 10.1109/TPEL.2012.2212917

Yuvaraj, T., Devabalaji, K. R., Kumar, J. A., Thanikanti, S. B., & Nwulu, N. (2024). A comprehensive review and analysis of the allocation of electric vehicle charging stations in distribution networks. IEEE Access. https://ieeexplore.ieee.org/abstract/document/10380227

Zentani, A., Almaktoof, A., & Kahn, M. T. (2024). A Comprehensive Review of Developments in Electric Vehicles Fast Charging Technology. *Applied Sciences (Basel, Switzerland)*, *14*(11), 4728. https://www.mdpi.com/2076-3417/14/11/4728. DOI: 10.3390/app14114728

Zheng, Y., Dong, Z. Y., Xu, Y., Meng, K., Zhao, J. H., & Qiu, J. (2013). Electric vehicle battery charging/swap stations in distribution systems: Comparison study and optimal planning. *IEEE Transactions on Power Systems*, *29*(1), 221–229. https://ieeexplore.ieee.org/abstract/document/6590033. DOI: 10.1109/TPWRS.2013.2278852

Zhou, X., Lukic, S., Bhattacharya, S., & Huang, A. (2009, September). *Design and control of grid-connected converter in bi-directional battery charger for plug-in hybrid electric vehicle application. In 2009 IEEE Vehicle Power and Propulsion Conference*. IEEE., https://ieeexplore.ieee.org/abstract/document/5289691

Zhou, X., Wang, G., Lukic, S., Bhattacharya, S., & Huang, A. (2009, September). *Multi-function bi-directional battery charger for plug-in hybrid electric vehicle application. In 2009 IEEE energy conversion Congress and exposition*. IEEE., https://ieeexplore.ieee.org/abstract/document/5316226

Chapter 3

A Survey on Efficient Coil Structures of Inductive Power Transfer for Charging Electric Vehicles (EVS)

H. D. T. Sangeetha
https://orcid.org/0000-0002-0073-8107
Karpaga Vinayaga College, India

L. Rajesh
https://orcid.org/0009-0008-2296-5723
SRM Institute of Science and Technology, India

Delshi H. D. H. Devi
https://orcid.org/0000-0002-7025-1193
Karpaga Vinayaga College of Engineering and Technology, India

H. D. K. Rakesh
https://orcid.org/0000-0002-7025-1193
Karpaga Vinayaga College, India

ABSTRACT

The world is being seriously affected by global warming, which is largely due to greenhouse gas emissions due to carbon emissions. However, 25% of all carbon emissions come from burning fossil fuels for transportation. Electrical vehicles have been chosen as an alternative to conventional to reduce carbon emissions to some extent. A charging technique for electrical vehicles plays a vital role in developing

DOI: 10.4018/979-8-3373-6335-6.ch003

Copyright © 2026, IGI Global Scientific Publishing. Copying or distributing in print or electronic forms without written permission of IGI Global Scientific Publishing is prohibited.

electrical and autonomous vehicles. As it has a number of advantages, a wireless power transformation has been the best option for charging a vehicle as it supports charging on the fly. The core of wireless charging is its coil structure and shape as an Inductive Power Transfer (IPT) is the most popular. The overall efficiency majorly depends on the coil and its parameters in addition to several control techniques for a larger power transfer. In this paper, a detailed study of various coil structures and shapes has been studied. Its efficiency and structure suitability are also discussed to assist the choosing a proper coil for the proposed wireless power transfer.

1. INTRODUCTION

Global warming, cloud bursts, the El Nino effect, acid rain, and climate change, which are brought on by the pervasive usage of vehicles with Internal Combustion Engines (ICE), are major concerns of the international community. Alternatives are thus needed, such as EVs that run on energy from renewable sources to reduce the dependency on sources of energy derived from fossil fuels and minimize their harmful effects on the planet Feng et al., (2021), Un-Noor et al., (2017), Ehsani et al., (2010).

Imagine a day in the future when an Electric Vehicle (EV) that is not being driven by a person needs to be charged on its own. In this case, a fully automated, rapid, safe, cheap, and dependable charging infrastructure is required to enable a profitable business model and the rapid uptake of electric vehicle systems. Current Wireless Power Transfer (WPT) Systems the advantages of technology include ease, safety, flexibility, and the possibility of total automation. The majority of cars using lithium-ion batteries nowadays are EVs. As the batteries' cruise range increases, the car will become heavier and more expensive. Some authors offered quick battery charging methods that might cut the total charging time to under 30 minutes Musavi et al., (2014). There are presently fast charging options, however, they are pricey and challenging to use. However, refuelling a fossil fuel-powered car takes less time than charging a battery does. Utilizing "swapping stations," which are areas where depleted batteries for electric vehicles can be exchanged for fully charged ones, is another option Feng et al., (2021). The utilization of charging systems is the primary element to drive the growth of EVs. The two types of EV battery charging methods that are now available are plug-in charging, also known as conductive charging or wire charging, and wireless charging, also known as inductive charging. The plugin charging system is further separated into On-Board and Off-Board chargers based on the charging infrastructure.

The fact that handling high-voltage wires for EV plugs can be challenging is one of the primary issues with conductive charging. Hazards may come from damaged cables or inappropriate handling. Techniques for conducting charges are likewise susceptible to theft and vandalism. The world's automakers and research organizations have been concentrating on Wireless Power Transfer (WPT) technologies since 2007 Emadi et al., (2003). Because of its high effectiveness and minimal servicing requirements, resonant inductive wireless charging should be given more importance in WPT approaches than other wireless power transfer systems. Nikola Tesla first proposed WPT, a different modern technology, in the 19th century. This technology developed over time and is now a competitive alternative to cable charging solutions. Electrical converters that convert electricity are used by the receiver in WPT systems to transmit power to the driving system or stored in batteries.

Electric automobile charging can be classified into three different types Yilmaz et al., (2013): i) Static Charging (SC) and Static Wireless Charging Method (SWC). ii) Dynamic Wireless Charging (DWC) Method where the vehicle is moving while charging. Another type is a hybrid type of iii) Dynamic Combined with Static Wireless Charging (QDWC) Method.

Static methods where the vehicle and charging pad are fixed which has the types of Static Plug-in Charging Method (SPC), There are four different Electric Vehicle (EV) charging methods are in charging methods Mode 1, Mode 2, Mode 3, and Mode 4. Categories are created depending on the battery's performance and charging time Etacheri et al., (2011), Tuttle et al., (2012). Mode 3 and Mode 4 fully charge a battery in less than 30 minutes and often need a three-phase supply, modes 3 and 4 are referred to as rapid charging Alanazi, (2023), Zhang et al., (2021). However, there are special features and limitations associated with each of these charging methods, which are described in Figure 1. Mojumder et al., (2022).

Wireless Power Transfer (WPT), is a method of recharging employing resonant coupled inductors Ahmad et al., (2022), Kan et al., (2018), Abdolmaleki et al., (2019), Gao et al., (2017). Many companies are manufacturing their static charging systems, which charge a car while it is still and maintain a specific degree of misalignment Yin et al., (2013), Patil et al., (2018). Many authors have put forth several designs for power electronics both on the surface as well as inside of vehicles. Ahmad et al., (2018), Kan et al., (2016), Mohamed et al., (2020), Laporte et al., (2019). For example, the author of Abdolmaleki et al., (2019) offers a 1 kW static charging device with 20 kHz rated frequencies that is based on adjustable frequency control. The 7-kW system that the authors of Miller et al., (2015), Rosu et al., (2016), Pevere et al., (2014), Petersen et al., (2014) discuss operates at its rated frequency of 79 kHz and has a fixed frequency control. A functional spectrum of frequencies and a comprehensive categorization of WPT systems based on the gap between the coils and the maximum allowable power level are provided by the SAE J2954 Recommended Practice (RP) Madawala et al., (2013), Deng et al., (2014), Hossain et al., (2022). The wireless power transfer is shown in Figure.2. On the base (or transmitter) portion of the entire system, a DC/AC converter, often an H-bridge, powers the transmitting coil and adjusts the parts. In addition, this converter serves as the Power Factor Corrector (PFC) for the system. On the vehicle (or receiver) side, a rectifier with a diode and the vehicle's battery is connected to the receiver coil and its adjusting components. An alternative DC/DC converter may manage the battery connection Zhang et al., (2015), although it is not taken into consideration in this study. Furthermore, various coil configurations and reference designs that operate as testing stands for simultaneous characterization and interconnectivity evaluation are provided by the SAE J2954 RP Liu et al., (2017), Villa et al., (2009), Azad et al., (2019), Sinha et al., (2019), Zhang et al., (2016), Mohamed et al., (2017), Cota et al., (2019), Ibrahim et al., (2015), Bucher & Duerbaum, (2012). Recently, advancements in Dynamic Wireless Power Transfer (DWPT) technology have made it feasible to increase the range that operates Electric Vehicles (EVs). The Features of Charging Technologies is shown in Figure 1.

Figure 1. The features of charging technologies

Mode 1	Mode 2	Modes 3 & 4

Slow Charging	Medium Fast Charging	Ultra-Fast Charging
Home charging	Parking Places and shopping malls	Public Charging Station
7-15 Hours	3-6 Hours	20-40 Minutes
63A Max Current	32A Max Current	16A Max Current

Figure 2. Schematic representation of the SAE 12954 RP-described power components of WPT systems. The analysis's objects are encircled by green dashed lines. Inspired by Zhang et al., (2015).

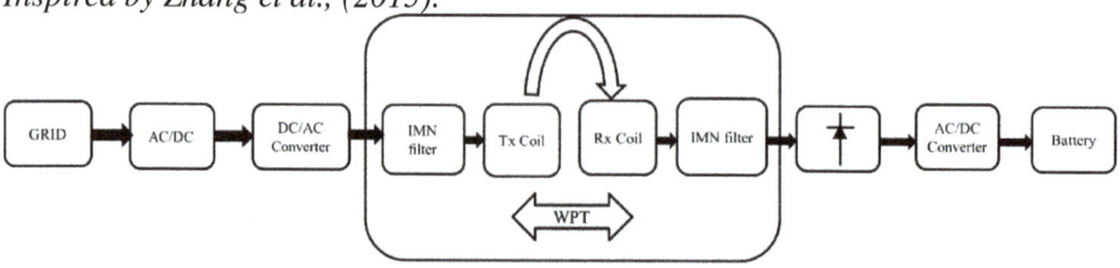

Inductive Power Transfer System (IPTS) in WPT operates on a similar concept to transformers, which also rely on electromagnetic induction. But IPTS and transformers are not the same thing; in the latter, there is strong coil coupling Zhang et al., (2021), Zhang et al., (2016). One of the main advantages of IPTS is the capacity for charging an Electric Vehicle (EV) when it is in motion. By reducing the size of the EV's battery and increasing its range of operation, this method has the potential to drastically change how electrical mobility advances in the future Jeong et al., (2018), Tang et al., (2015). The efficiency of the Wireless Power Transfer System mainly depends on coil structure, phase matching between coils, and load-dependent on the secondary coil, given the input load of the primary coil Mou et al., (2020), Kim et al., (2016), Chowdhury et al., (2019). As we are going to give importance to the coil on the efficiency, the efficiency of the coil depends on the various coil factors such as Magnetic Structure in the Power Pad, Coil structure, Airgap between the coils, coil coupling technique, Magnetic resonant coupling, number of Coil turns coil weight, Impedance matching between the coil with various speeds, etc Liu et al., (2013). Schematic representation of the SAE 12954 RP is shown in Figure 2.

In the next section, the Magnetic structure of a power pad is discussed. Further, coil structures and their suitability for electrical vehicles in terms of efficiency have been provided in Section 3. In section 4, the effect of coil parameters such as air gap, coil coupling, magnetic resonant coupling, coil turns, and coil weight on the efficiency have been discussed. Finally, how the speed of the car would affect the efficiency in the dynamic wireless charging has been detailed and coil structures are compared on their efficiencies. The paper is concluded in the last section.

2. LITERATURE SURVEY

Emrullah Aydin et.al Nowadays, Wireless Power Transfer (WPT) technology receives increasing attention in the automotive sector, introducing a safe, flexible, and promising alternative to standard battery chargers. Considering these advantages, charging electric vehicle (EV) batteries using the WPT method can be an important alternative to plug-in charging systems. This paper focuses on the Inductive Power Transfer (IPT) method, which relies on the magnetic coupling of coils to exchange power from a stationary primary unit to a secondary system onboard the EV.A comprehensive review is performed on the history of the evolution, working principles and phenomena, design considerations, control methods, and health issues of IPT systems, especially those used in EV charging. In particular, the coil design, operating frequency selection, efficiency values, and the preferred compensation topologies in the literature are discussed. Published guidelines and reports studying the effects of WPT systems on human health are also provided. Additionally, suggested methods in the literature for protection from exposure are reviewed. The control section presents the common charging control techniques, focusing on the constant current-constant voltage (CC-CV) approach, which is typically used for EV battery chargers.

Aganti Mahesh et.al Considering a future scenario in which a driverless Electric Vehicle (EV) requires an automatic charging system without human intervention, there is a demand for a fully automatable, fast, safe, cost-effective, and reliable charging infrastructure that enables a profitable business model and promotes fast adoption in electrified transportation systems. These qualities can be achieved through wireless charging systems. Wireless Power Transfer (WPT) is a futuristic technology offering flexibility, convenience, safety, and the potential for full automation. Among WPT methods, resonant inductive wireless charging gains more attention compared to other methods due to its high efficiency and easy maintenance. This literature provides a review of the status of Resonant Inductive Wireless Power Transfer Charging technology while also highlighting the current status and future prospects of the wireless EV market. First, the paper delivers a brief history that sheds light on wireless charging methods, emphasizing their pros and cons. Next, it presents a comparative review of different types of inductive pads, rails, and compensation technologies. The static and dynamic charging techniques and their characteristics are also illustrated. The role and significance of power electronics and converter types used in various applications are discussed. Additionally, the batteries and their management systems, along with various challenges involved in WPT, are addressed. Topics such as cybersecurity, economic effects, health and safety, foreign object detection, and the impact on the distribution grid are explored. Prospects and challenges associated with wireless charging systems are also highlighted in this work. We believe this study can support further research and development of WPT systems.

Sampath Jayalath et.al The magnetic coupler serves as the heart of an inductive power transfer (IPT) system, enabling wireless power transfer through its air gap. The couplers are designed to maximize efficiency, power density, power transfer distance, and misalignment tolerance while minimizing leakage flux, weight, cost, and volume. The coupler design process becomes complex due to the nonlinear behaviour of magnetics, sophisticated geometrical structures, and mandatory design limitations imposed by standards such as SAE J2954/1, IEC 61980-1:2015, and ISO 19363:2020.This article reviews advancements in coil design methodologies and their structures over the past few decades to identify ongoing challenges and trends. It identifies the impacts of the power electronics system, industrial standards, material selection, numerical and analytical modelling methods, and thermal modelling on the coil design process to formalize the design procedure. A coil design example based on finite element analysis (FEA) tools is presented to illustrate the drawbacks of existing design and optimization processes. A sensitivity analysis, 3-D Pareto plots, and optimal design selection considering misalignment variations are proposed to improve the Multi objective optimization process. A generalized guideline for coil design is outlined, highlighting the essential design stages of an IPT coupler. Current trends are identified, and future directions are proposed.

Saeid Ghazizadeh et.al Given the substantial role of electric vehicles (EVs) in decarbonization, the demand for EVs becomes increasingly widespread. However, batteries remain a major challenge, primarily due to critical environmental concerns. Inductive Power Transfer (IPT) emerges as an effective technique to reduce reliance on batteries by enabling charging without physical connection. This paper provides a thorough analytical review of the latest advancements in the design of IPT couplers and identifies current gaps and future opportunities. Different aspects of IPT systems are reviewed, various coil designs are compared, and their performance is analyzed. In-depth theoretical analysis is performed through mathematical modelling, with electric and magnetic equations derived to ensure accurate outcomes. Furthermore, a magnetic field calculation tool is developed using a Python script to perform magnetic computations and generate 3D graphs of the total magnetic flux linkage. Additionally, performance evaluations for complex couplers are performed based on simulation results executed in Ansys Electronics software. To ensure comprehensiveness, trade-off trends in the design of magnetic couplers are presented, and relevant optimization techniques are reviewed and explained. This paper is intended to serve as a comprehensive reference for researchers seeking both in-depth theoretical analysis of IPT systems and an extensive review of coupler designs.

Geetha Palani et.al Considering a scenario where a driverless electric vehicle (EV) requires an automatic charging system without human intervention, there is a pressing need for a fully automated, fast, safe, cost-effective, and reliable charging infrastructure. Such an infrastructure must accommodate a range of EVs, be profitable, and enable rapid adoption in electric transportation systems. Wireless charging methods have the potential to meet these requirements. Wireless Power Transfer (WPT) represents a futuristic technology offering flexibility, convenience, safety, and full automation capabilities. Among WPT methods, resonant inductive wireless charging stands out due to its high efficiency and ease of maintenance, deserving more focus compared to other techniques. This literature provides an overview of the current status and future prospects of Resonant Inductive Wireless Power Transfer Charging technology, along with insights into the wireless EV industry's evolution. Initially, the article reviews the history of wireless charging technologies, highlighting their benefits and drawbacks. It then presents a comparative analysis of various WPT methodologies, control strategies, and compensation networks studied to date. Static and dynamic charging methods, along with their characteristics, are illustrated. A novel approach involving superconducting materials in coil designs is explored to enhance charging power, with emphasis on their potential impact on wireless charging systems. The detailed role of power electronics and the various types of converters used in different applications are discussed. Additionally, battery management systems and the challenges faced in WPT implementations are analyzed. Broader concerns, such as cybersecurity, economic implications, health and safety, foreign object detection, and the effects on the distribution grid, are investigated. This article also examines the opportunities and challenges associated with wireless charging systems. We anticipate that this work will contribute significantly to advancing research and development in WPT systems, facilitating their practical application in the EV industry.

3. MAGNETIC STRUCTURE IN POWER PAD (POP)

The electromagnetic coil that creates and links the magnetic field is located inside the power pad (PoP). The front of the pad is the side that faces the opposing coil; the backside of the pad is referred to in the same way. The fundamental parts of a PoP are illustrated in Figure.3. The positioning of the coil determines the type of bonding and flux production in the PoP. To create the voltage in the secondary coil, just one flux needs to flow across the center of the coil and parallel to its axis Zhang et al., (2016). When both the primary and secondary pad centers are facing one another, ideal alignment occurs, which is said to be the most efficient configuration. The secondary cannot take in all of the entering flux from the main pad, therefore if they are not aligned, the mutual coupling coefficient (MCC) lowers. As a result, leakage flow rises even more, eventually lowering the effectiveness of the system. The leakage fluxes are described as a flux that distributes randomly and does not connect the main coil with the secondary, which can be harmful to human health if it is over a certain level. International Commission on Non-Ionizing Radiation Protection (ICNIRP) has established the amount of leakage flux threshold at 27 μT for the operational frequencies ranging from 10 kHz to 100 kHz Tejeda et al., (2016), Zhang et al., (2015). The structure of ferrite has been modified to reduce the leakage flux. It is one of the most important things to consider while developing the PoP. The flux leakage is controlled by the aluminium rings and plates Kim et al., (2016). Usually, the coil is placed on a ferrite that extends from copper.; This ferrite provides a route of least resistance to the flux, therefore modelling it, directing it in the direction of the pad's face rather than going down to the earth. Lin et al., (2015) Aluminium base provides a base support for the coil. However, a portion of the flux interacts with this aluminium. It creates eddy currents that circulate within it, lowering the coil's Q and resulting in power losses. Due to its greater voltage-ampere (VA) evaluation, the main pad is the main cause of leaky flux.; The primary pad is the major source of leakage flux because of its higher VA rating; initially, less flux is caught, and the rest of the amount is classified as leakage. Additional ferrite has been exposed in the layout of the Circular Pad (CP) and Bipolar Pad (BPP) primary pads to reduce leakage flux at the cost of MCC and Q Lin et al., (2014). The created PoP may be used in a variety of airgap applications; because of this, the secondary coil, which is positioned away from the primary coil, needs to be coupled by the flux it creates spreading out over enough time. Among the most important components of a well-developed PoP is compatibility. These kinds of pads aren't compatible with charging stations Thus, the need for a pad to be interoperable makes sense from a business perspective. The mutual coupling coefficient (MCC) of PoP is used in the literature to compare them., Within the aforementioned requirements are their capacity to transmit power wirelessly, their efficiency with lateral misalignment, and their suitability with different types of pads are taken into consideration. Budhia et al., (2011). Inner Structure of Power Pad is shown in Figure 3.

Figure 3. Inner structure of power pad

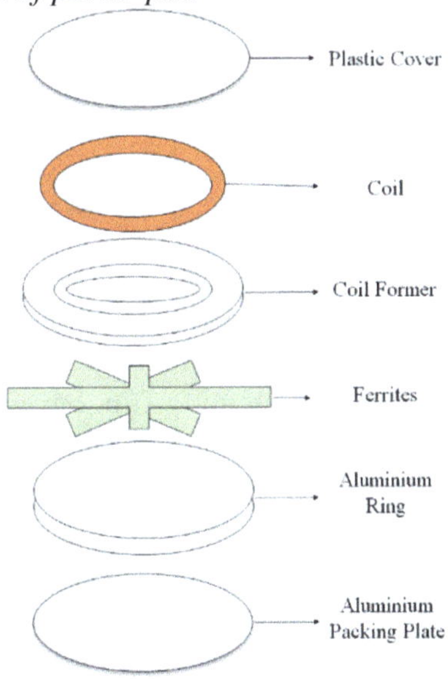

4. COIL TOPOLOGIES

Based on the flow direction of flux generation, interaction, or susceptibility, magnetic coil builds, coil layouts, and power pads may be divided into three primary groups: polarized pads (PP), non-polarized pads (NPP), and both Lin et al., (2014).

Non-polarized pads (NPP) produce perpendicular flux, including circular pads (although square and rectangular geometries are also possible), and are vulnerable to it. The PP creates a flux that is either parallel or horizontal and is coupled to the pad layer Budhia et al., (2011). Because, depending on their phase currents, they may create both types of flux patterns, Within the PP & NPP group are the Tripolar Quadrature (TQP), Double-D Quadrature (DDQP), Tripolar (TPP), Bipolar (BPP), and Double-D Circular (DDC). Figure.3 shows the power pad classification tree based on the coil structures or topologies. Classification of Coils based on Structures is shown in Figure 4.

Figure 4. Classification of coils based on structures

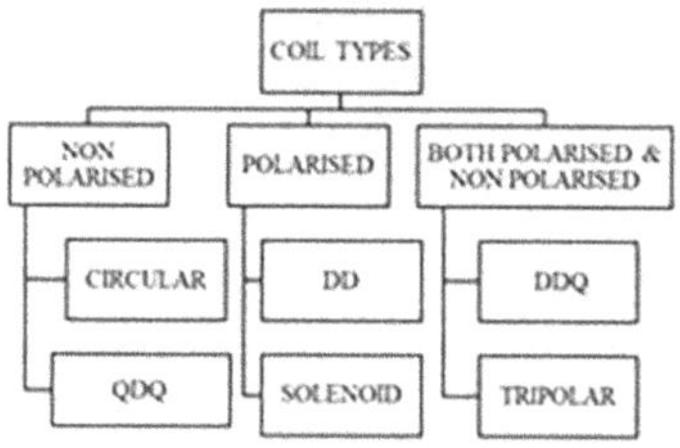

4.1 Circular Pad (CP)

The CP design depicted in Figure. 5.a has been examined and improved in Ahmad et al., (2018). A circular inductor coil is positioned above ferrite layers to enhance the electromagnetic field pattern by rerouting flux out from the pad. The gadget is small and easy to assemble. As can be seen in Figure.5. b., an aluminium plate and ring are also employed to provide shielding and control flux leakage from the bottom of the pad. Given that it only blocks leakage flux and not the main flux, it has no appreciable impact on the pad's efficiency. Aluminium must be placed to the rear of the EV to protect it from leaking flux, which might gather in the steel floor plane and create losses. Lin et al., (2015)

Figure 5. a) Structure of circular coil; b) Pad arrangement of CP; c) Magnetic flux of the circular coil; d) Voltage reduction as horizontal misalignment of CP

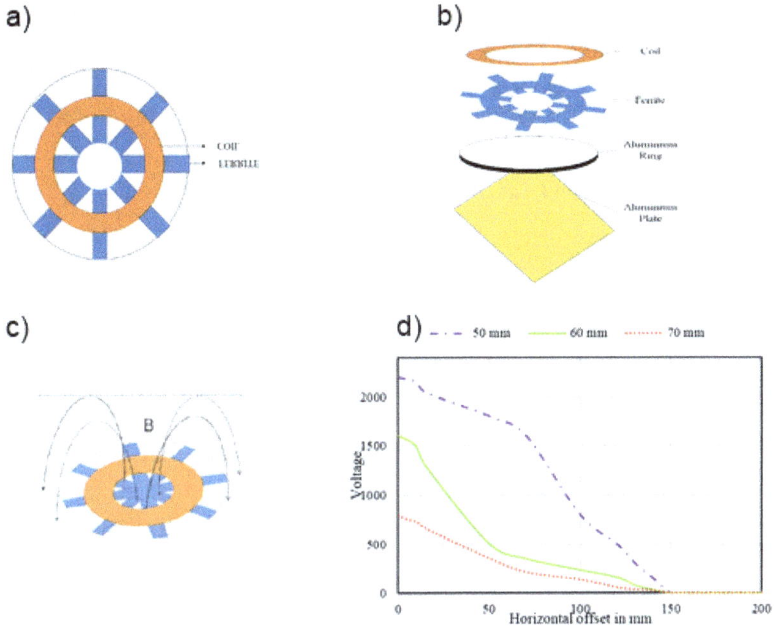

Because of the presence of electromagnetic distortion, circular-shaped coils have a greater mutual coupling coefficient (MCC) than rectangular designs or square-shaped design coils. As can be observed in Figure. 5.c, circular flux disperses in all directions, which is an undesirable property of CP and lowers the MCC Tejeda et al., (2016) as well as the leaky flux Zaheer et al., (2014). It is being shown that, when considering both vertical and horizontal displacements from central coordination, 90 kHz is the ideal frequency for CP coil performance. Because its field propagates in all directions, CP is not affected by secondary pads such as BPP rotating. The major CP limitation that results from pads having the power profile and magnetic coupling are both null when there is horizontal misalignment Ahmad et al., (2018).

Currently, there is no magnetic interaction between the two pads, therefore the system's power transmission efficiency and effectiveness are either very low or non-existent. It happens as a result of the principal pad's opposing flux density vectors canceling each other out. Additionally, it appears when the horizontal separation between both pads is about 38% of the pad's diameter. The air gap across the coils has no effect on the null point, which is evident at 160 mm of the horizontal shift from the centre of the pads, each of which has an outer diameter of 420 mm, versus varied air gaps as explained in Figure.5. d.

Although not optimal for large airgap applications, CP has been widely used in a range of pads. Relative to other coils, it has a simpler design, is less expensive, and leaks less flux Budhia et al., (2011). Under high air gap applications, the performance metrics of CP degrade. Lin et al., (2014), Budhia et al., (2011). The pad is larger and unsuitable for powerful and air gap usage as its diameter has to expand four times to provide the same quantity of power. Lin et al., (2015) Bosshard et al., (2013).

4.2. Design of Quad-D Quadrature Pad (QDQP)

The graphic displays a novel coil layout known as Quad-D quadrature (QDQ). This pad is composed of a bigger quadrature coil, as seen in Figure 6.a, which is constructed from four more circular coils stacked one on top of the other. The aforementioned coils are connected electrically together in parallel but magnetically in sequence. Tests have been conducted on the recommended pad at a range of air gaps, from 50 mm to 250 mm. In summary, it exhibits excellent MCC and deteriorates with increasing air gap width. Along with adding to the airgap, an effective lateral misalignment of half of the pad's size has also been attempted. This system is 91.4% efficient when it is properly aligned, but only 78% efficient when it is at its maximum offset Chowdhury et al., (2019).

Figure 6. a) Structure of quad-D quadrature coil; b) Structure of solenoid pad

a)

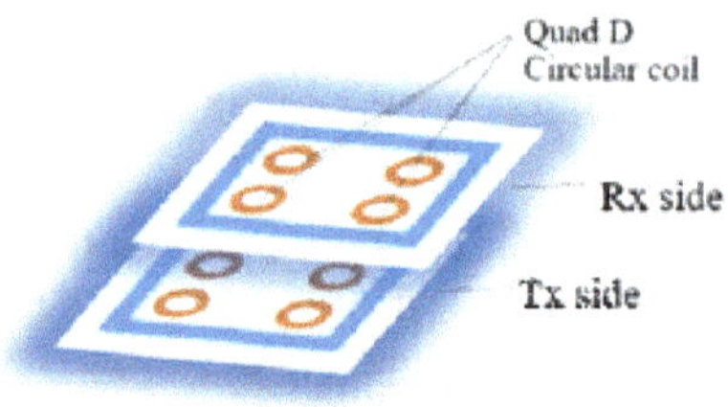

b)

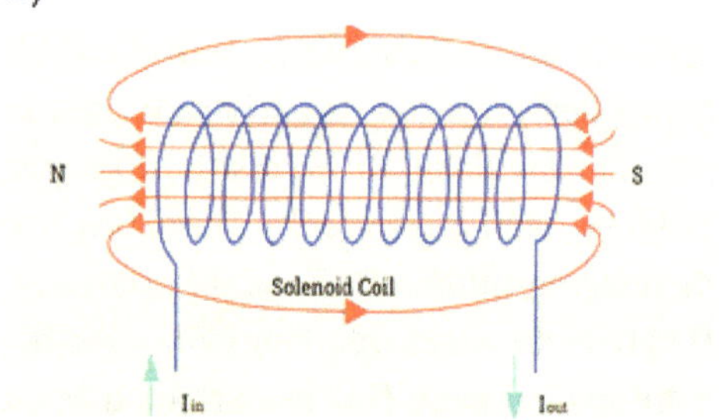

4.3 Solenoid Pad (SP)

The solenoid pad (SP) produces double-sided opposing polarity flux and is a member of the polarized flux class as given in Figure.6. b. Zhang et al., (2015). It has been used for IPT because of its straightforward construction, especially in low-power applications. Made of ferrite and litz wire, as well as it is a vertical coil. The power pad for charging electric vehicles has to be compact enough to be set on the ground. If the solenoid assembly is positioned horizontally on a flat surface, the flux direction will be horizontal, and it cannot be connected to the supplementary pad of the vehicle, which is situated vertically.

While SP can work with both main and secondary pad types, it has several undesired characteristics, such as producing double-sided f lux and a lower Q. Solenoids, also aren't appropriate for use in primary pads for electric vehicle (EV) charging systems because of their high leakage flux. By making the ferrite thicker, these systems' poor magnetic efficiency may be raised. Zhang et al., (2015), Budhia et al., (2011), as well as Mosammam et al., (2019).

4.4 Double-D Pad (DDP)

DDP was first developed by Kalwar et al., (2016) in response to the drawbacks of CP, including a narrow flux path, low coupling factor, and displacement intolerance. It consists of two rectangular-shaped co-planar coils positioned on ferrite bars that run along the length of the pad. These coils are twisted to mimic the number eight, as seen in Figure.7. a. If it is utilized as a main pad, it is frequently coiled in the form of a coil with two terminals. DD coils have four terminals and can be powered by two distinct power sources and coiled independently. Budhia et al., (2011). Because those two coils are powered by sources that supply out of sync current, the current in one coil flows clockwise, while the current in the additional coil flows anticlockwise. It frequently appears in the primary pad. It is crucial to keep in mind that both of these coils are now connected.

Figure 7. a) Structure of the double-D pad coil; Structure of the double-D quadrature pad coil

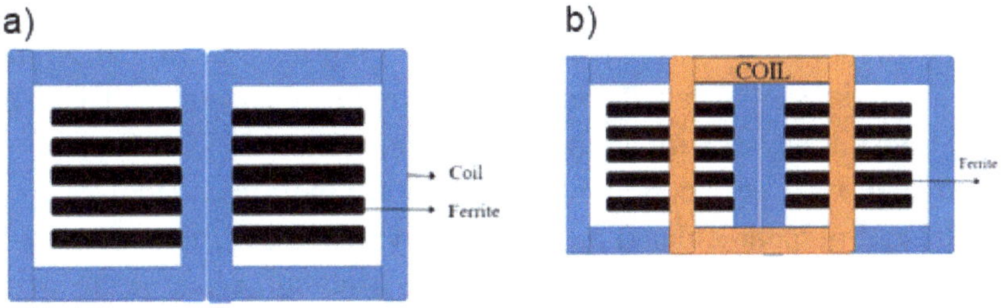

To minimize copper losses, other coil components should have the shortest possible lengths. In contrast, the flux pipe generated by joining the two coils should have the longest possible length for design considerations. Lin et al., (2015). The term "Double-D" refers to the flux that is produced when two coils are put next to each other and magnetically connected. This flux has a shape like a "D". We generate and gather the unidirectional flux parallel to the DDP's surface. It generates two opposites. Budhia et al., (2011). The flux travels down the pad's length, resulting in a polarized horizontal flux profile. Due to the minimal leaky flux and loss brought on by the intended one-sided flux arrangement, "Q" is not particularly low when aluminium is positioned towards the rear of the pad. Lin et al., (2015), Kalwar et al., (2016). DDP is frequently supplied by a single power source because both of its terminals are reachable from the pad Budhia et al., (2011). DDP has an MCC that is between twenty percent and twenty-five percent more than CP's. It is advised to utilize the DDP architecture for dynamically charging applications since, at the same cost and size, its charging area is five times bigger than the CP's. The charging zone of DD concerning CP is shown in Figure.7.b. The flux path height in DDP topology is double as long as it is in CP Kim et al., (2016), Budhia et al., (2011), Budhia et al., (2011), Shimizu et al., (2013).

DDP exhibits minimum in magnetic and power profiles down the horizontal axis, much like CP. When the DDP is thirty-four percent of the pad's length off-cantered around the other coil in the horizontal direction, the power profile is nil. Lin et al., (2016). Its absence of the null along the y-axis, qualifies it for application in dynamic charging systems. Covic et al., (2013), Jamakani et al., (2019). DDP produces a denser field with greater MCC than SP, although it has less leakage flux than CP.

The following are the main benefits of DDP that make it suitable to use for large airgap scenarios: greater flux pathway height, which is proportionate to fifty percent of pad length rather than a quarter of pad width as in CP; decreased leakage flux as a result of a one-sided flux pattern; Efficiency can tolerate a bigger charging zone than CP and aluminium shielding at the rear. In addition, compared to CP, DDP has enhanced Q. DDP may be applied to dynamic as well as static charging applications due to its broad and dense flux coverage along x, y, and z (great air gap). A comparison between CP and DDP may be seen in Covic et al., (2013). After testing the two coils under different vertical mismatch conditions, it was determined that DDP had a higher MCC than CP. It implies that with a mismatch and a big air gap, DDP may transmit power with greater effectiveness than CP

CP and DDP have very poor compatibility since their intended uses are to collect parallel and perpendicular flux, respectively. Since both centrally aligned CP and DDP are intended to collect parallel and perpendicular flux, respectively, this is the primary cause of DDP's issues and the reason it doesn't function well with CP Jamakani et al., (2019). It is advised to misplace the secondary NPP coil in this situation to avoid the zero, even if doing so reduces efficiency and necessitates flux leakage charging applications Panchal et al., (2018).

4.5 Double-D Quadrature Pad (DDQP)

The absence of the null point limitation in the power profiles of DD and CP pads further encouraged the researchers to design multi-coil pads that would connect parallel and perpendicular flux to avoid null but also be more lateral distortion intolerant than prior ones. A horizontal flux-producing DD coil plus a quadrature coil that generates some one-sided vertical flux make up a Double-D Quadrature pad. As seen in Figure.6.b, the ferrite is also positioned behind the quadrature coil and extends somewhat past the DD coil to improve the coupling even further. Reachable DDQP terminals are four. With the axes of both coils parallel to one another, the quadrature coil and DD coil fit in the centre of the pad below one another. Because of the way they are positioned, both of these coils are mutually decoupled, a highly desirable property of a pad that allows for individual adjustment and control from two different power sources. Terminals are devices that are physically reachable and not part of the DDQP and may be subject to separate regulations. For every coil, there are a pair of connections. In DDQP, out-of-sync currents are used to energize the Double-D coil to create or couple.

The pad's length is generated by a polarised parallel magnetic field, while non-polarized perpendicular flux is produced by the grid. The fact that the field in a quadrature coil is not perfectly symmetrical at the pad's center distinguishes its flux pattern from that of a circular coil. By varying both the phases and magnitudes of the main current, the flux form of a system with two additional coils driven by DDQP may be altered.

Compared to CP, the charging area of DDQP is approximately three times larger. The larger charging zone of the DDQP makes it a desirable option for universally compatible pads with a smaller overall dimension and the same amount of material utilized, it has been shown that the DDP-DDQP system has a charging area that is around three times greater than the DDP-DDP system and five times greater than the CP-CP system. Budhia et al., (2011) Covic et al., (2010).

But based on the needs, it can only produce parallel fluxes by activating the DD coil, perpendicular flux by energizing the quadrature coil, or a combination of fluxes. The primary benefit of DDQP is in its ability to generate flux in both parallel and perpendicular directions, hence enhancing its misalignment intolerance and charging zone properties. Budhia et al., (2011).

Due to the additional coil that improves interoperable by producing a perpendicular flux, DDQP is more costly than DD and CP. The weight of this pad is a further significant drawback, in addition to the higher copper content. The DDQP also enhances the pad's total dimension since it combines the DD and Q pads.

4.6 Tripolar Pad (TPP)

Recently, a multi-coil pad called the Tripolar pad (TPP) was released by Zhang et al., (2021). Instead of the strips of previous pads and aluminium plates, it is placed above hard ferrite and is made up of three partially overlapping coils, as depicted in Figure 8. a & b. Similar to BPP, these coils are meant to be mutually disengaged. Nonetheless, there is a cross-coupling between the coils in the event of a misaligned secondary. Once mutual decoupling has been achieved as previously described, the magnetic field's form may be adjusted based on the secondary pad by adjusting the phases and magnitudes of the primary current in each coil. The TPP generates both horizontal and perpendicular fluxes, similar to BPP, circular, triangle, and hexagonal as given by the following equations.

Figure 8. a) Basic structure of the tripolar pad; b) Structure of the tripolar pad coil

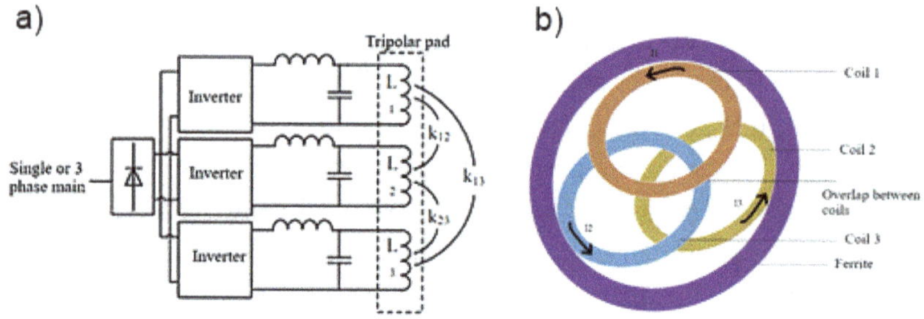

$$P_o = V_{oc} I_{sc} Q_2 = \omega L_p I_1^2 k^2 Q_2 V_{oc_n} = \sum_{m=1}^{M} j\omega k_{mn} \sqrt{L_m L_n} I_m \angle \theta_m \quad (1)$$

$$I_{sc_n} = \sum_{m=1}^{M} k_{mn} \frac{\sqrt{L_m}}{\sqrt{L_n}} I_m \angle \theta_m \quad (2)$$

$$P_o = \sum_{n=1}^{N} V_{oc_n} I_{sc_n} Q_2 \quad (3)$$

The entire number of mutually decoupled coils in the main pad and supplementary pad, denoted by M and N, accordingly, is expressed by the coupling factor k_{mn}, which connects coil m in the main pad and coil n in the supplementary pad. The phase discrepancies between the principal currents are denoted by the letter m. The voltage that causes an open circuit and short-circuit current generated in coil 'n' of the secondary pad are denoted by the symbols Voc_n and Isc_n, correspondingly. Voc_n

is the total voltage produced in the coil n by all of the coils in the primary pad for the phase differences, considering the concept of superposition.

The neighboring coil of a TPP experiences an electromagnetic field (EMF) due to the partially interconnected segment of the powered coil and the opposite EMF produced by the rest of the coil. Each of the three coils effectively alter their overlap, to minimise the net EMF generated in an adjacent coil. This allows for simultaneous decoupling between the three coils.

Finally, the comparison of all structures is given in Table 2. The misalignment of the coils affects the secondary voltage reception as shown in Figure 9. From the comparison chart, it is evident that the triple Q and DDQ pads offer low loss, and hence their secondary voltage reception is also high as compared with other structures.

Table 2. Comparison of coil structure

Coil Type	Type of flux (Polarised & Non-Polarised)	Misalignment	The Collaboration	Comparative Material Cost
Circular	Non-Polarised	extremely low	extremely low	low
DD	Polarised	low	extremely low	low
DDQ	Both	moderate	higher	higher
Quad D	Non-Polarised	higher	higher	moderate
Tripod	Both	Very high	Very high	moderate

Figure 9. Comparison of transmitter-receive voltage with the uncoupling coils for different structures.

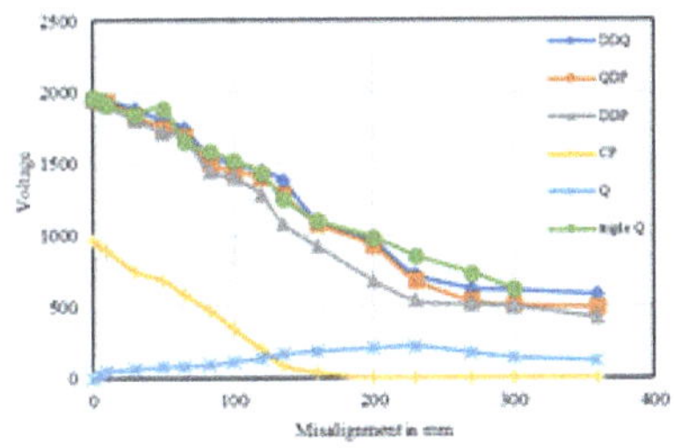

5. EFFECT OF OTHER COIL PARAMETERS

5.1. Air Gap Between Coil

Due to the attenuation of the field within the coils, the inter-layer inductance that exists reduces as the distance, d, grows. In this case, an increase in radius causes less inductance. This is mostly because, for the fields created by the turns to join with those created by surrounding turns, they must travel farther. At this distance, the attenuation effect becomes stronger. The aforementioned conduct appears in every narrative. Liu et al., (2017), Zaheer et al., (2012).

Several lamped coil designs energy transfer efficiencies are simulated and examined, along with the effects of various turn counts and inter-turn distances for transmitter and reception coils. The power transmission efficiency can be improved by increasing the number of turns on both coils and the distance between turns. Ramezani et al., (2021), New et al., (2019).

5.2. Coil Coupling Technique

When there is connectivity, the coil's power transmission capacity is unaffected by the shape and power of the receiving coil—a feature that is advantageous to automobiles. It should be possible for the secondary and primary coils to pair together successfully. Efficiently collecting the coil that transmits flux in both parallel and perpendicular field patterns must be the responsibility of the coil that receives the signal. Ali et al., (2022).

It is possible to construct a highly efficient loosely connected transformer using interconnection research. It is possible to construct transformers with loose connections to have the highest coefficient of coupling (k). Without misalignment or with the worst possible misalignment, the maximum k can be reached. According to the crucial pad design, a coil's capability should be independent of its symmetrical coil shape. The WPT system is proposed with electric and magnetic interconnection by international standard SAE J2954.

Impedance matching networks and power converters are essential for electrical interconnection. The size of the transmitter and receiver pad coils affects magnetic connection. Conversely, the lowest connection properties are seen in the DD pad coil and the circular pad coil. It will be shown as having zero-coupled power when one of the pads is positioned in the middle of the other. A DD coil generates horizontal flux, as opposed to vertical flux produced by a circular coil. The DDP coil and CP coil systems on a DDQ pad will function effectively altogether. Moreover, compared to non-polarized pads, it will offer a bigger charging zone. The best field pattern may usually be captured by the receiver using the DDQ and BP pads. As compared to other rectangle and DD coils, the quadrupole coil architecture has the greatest coupling coefficient with the DDQ coil. The additional quadrature coil in the quadrature coil accounts for its low k and M values Küpper et al., (2014). There is not a noticeable misalignment between the CP and DDP structures and a newly installed multi-coil ground arrangement Zaheer et al., (2015). Using the interface impedance method, decompose power A method for evaluating interconnection was proposed by SAE J2954. However, the approach is not suitable for practical use because it measures phase angle. An orientation-adjusting technique based on Varying Step Perturbation and Observation (VSP and O) was proposed Ahmad et al., (2019), Lin et al., (2020).

5.3. Magnetic Resonant Coupling (MRC)

Subcategories of Wireless Power Transfer (WPT) technology exist. WPT technologies may be divided into two groups: near-field WPT technology and far-field WPT technology, based on the power supply range. The electromagnetic radiation used by a solar satellite, laser, or microwave is the basis of the far-field WPT approach. WPT connections such as Magnetic Resonant Coupling (MRC) WPT and Inductive Power Transfer (IPT) WPT are the two varieties of near-field WPT technology. While IPT WPT is superior for high-voltage transfer of power since there is no resonant circuit involved, MRC WPT is particularly helpful for low- or medium-power WPT because both the primary and secondary coils may be modified in resonance frequency by adding compensating capacitors. The benefits of MRC WPT include its high level of safety and wide gearbox range. A comparison of the MRC WPT and IPT WPT is shown in Table 3.

Table 3. IPT WPT vs MRC WPT comparisons

	IPT WPT Non-Resonant Coupling	**MRC WPT Resonant Coupling**
Model Circuit	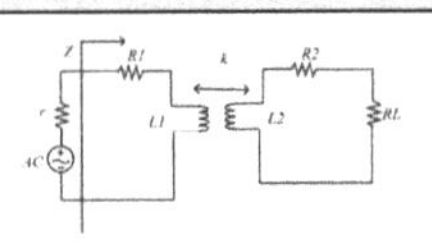	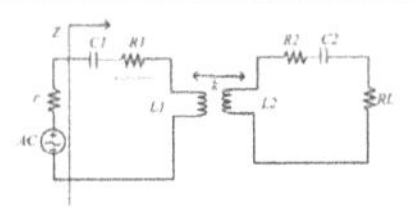
Equivalent Impedance Z	$Z = R_1 + j\omega L_1 + \frac{k^2\omega^2 L_1 L_2}{j\omega L_2 + R_2 + R_{Load}}$	$Z = R_1 + \frac{k^2\omega^2 L_1 L_2}{R_2 + R_{Load}}$
Merits	sample-safe, very efficient transfer	greater transmission range compared to inductive WPT
Demerits	short broadcast range.	adjusting the resonance frequency for various devices is difficult.

5.4. Coil Size Depending on Coil's Turns

The coils with the smallest coil installation rate, broken down by coil size, are shown in Figure.10 a, it shows the number of coils positioned every 100 kilometres as well as the average efficiency. Average efficiency somewhat decreases with coil size; However, there is a chance that coil installation rates may drop and that there won't be as many coils installed every 100 km. The benefit of cutting the coil installation cost will exceed the power loss throughout the course of the system's 30-year lifespan. Thus, while restricting the total number of coils inserted, it is crucial to take coil size L and coil installation rate into account. Yang et al., (2021).

Figure 10. a) Comparisons of increase the coil turns vs. Efficiency of the coil; b) Comparisons of coil weight vs. power transferred in terms of Voltage

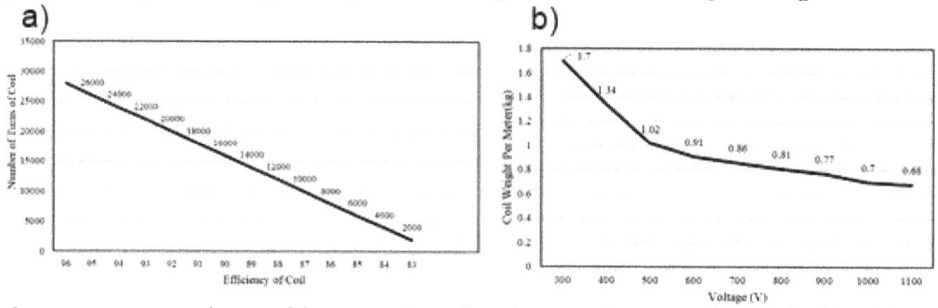

5.5. Weight of the Coil

Since copper is required for infrastructure, charging stations, and car interiors, it is an essential part of the electric transportation industry. The majority of an electric automobile, bus, or train's wiring, electric motor(s), and batteries are made of copper. A typical passenger electric vehicle (EV) uses around 80 kg of copper, or 4% of its total weight. An average battery-powered electric bus weighs around 370 kg in copper. DC link voltage has a significant effect on copper weight within an EV as shown in Figure.10. b. Due to tripling the battery voltage, Porsche claims that it has decreased the weight of its electrical harness by 30 kg Yang et al., (2021), Yamada et al., (2021). 1.2 kV rather than 400 V in a 150-kW system, as seen in Figure. 9.b., can lower the weight of copper per meter by 63% because the current is reduced from 375 A to 125 A. Additionally, higher voltage cables typically exhibit greater flexibility for routing due to the more than threefold reduction in wiring harness cross-section, which makes packing and manufacturing easier Yamada et al., (2021), Tan et al., (2020).

5.6. Primary and secondary coil impedance matching control with various Speed

The desired load value and coupling coefficient in DWC systems change as a function of the receiving coil's movement Meng et al., (2024), Küpper et al., (2015), Diep et al., (2019), Kobayashi et al., (2015), Hata et al., (2016), Dai et al., (2018).

Therefore, it is necessary to determine the appropriate load in advance to apply the impedance matching control. Put another way, the coupling coefficient must be continuously assessed. Some scholars have computed the e-coupling coefficient Chowdary et al., (2020). A real-time coupling coefficient estimation technique for a dynamic charging system is given in References Küpper et al., (2015), Diep et al., (2019), Kobayashi et al., (2015), Hata et al., (2016). For measurement, just the secondary side parameters are required.

The Root Means Square error (RMS) of the alternating current can be measured in order to account for errors in the discontinuous current mode. The DC side measurement characteristics can be used to determine the coupling coefficient Küpper et al., (2015), Dai et al., (2018), Jiwariyavej et al., (2015). To lessen power pulsation in the system, the magnetic coupler was built using FEA modelling Yin et al., (2016), Zhang et al., (2015), Diep et al., (2020).

5.7. Comparison of Coil Structures

Coil shapes on various parameters such as Tolerance of Misalignment, high EMF, and their applications are compared and listed in Table 4. The tripod offers a high coupling and good tolerance among other coil shapes. Further, structures are compared on parameters such as polarization, flux creation, and empty zone shown in table 5. It shows that a tripod structure offers a higher flux of two times of its sides.

Ansys Maxwell simulator is used to build a coil, and the structure that is constructed for the purpose of identifying the magnetic field that is produced by a single transmitter and a single receiver is also designed. Additionally, a multi-transmitter coil and a single receiver are designed. In addition, the total number of tur is two, and the input voltage is specified as ten volts. The design coil alignment between the coils is fifty millimetres, and the coil diameter is six millimetres.

We were able to achieve a more favourable magnetic field overall structure in the multi transmitter-single receiver configuration is 0.000095 B(Tesla)), in comparison to the single transmitter-single receiver configuration is 0.000049 B(Tesla)) shown in figure 11.

Figure 11. Comparison between the multi-transmitter and single transmitter with a single receiver

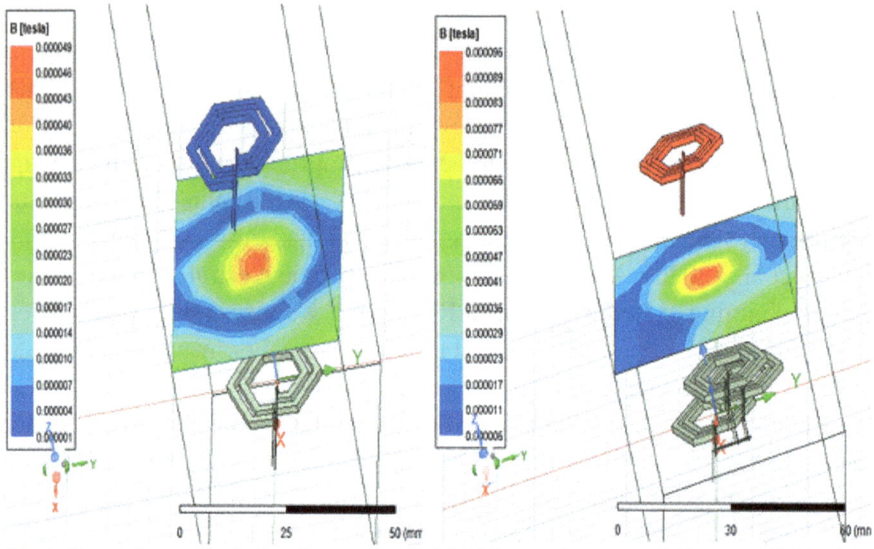

Single Transmitter – single Receiver (0.000049 B(Tesla))

Multi Transmitter – single Receiver (0.000095 B(Tesla))

Table 4. Comparison of coil function

Coil Shape	Tolerance of Misalignment	High EMF	Use
Circular	Limited range and null zones occur	Allow single-sided flux path and shielding	Vehicles approach the coil from various angles
DD	Suffers from null zones	Allow single-sided flux path and shielding	Operation in large air gaps
DDQ	Misalignment tolerance over the DD pad	Allow single-sided flux path and shielding	low misalignment
Quad D	Misalignment tolerance over the DD pad	High EMF	large charging area
Tripod	Allows high coupling factor and good tolerance	Allow single-sided flux path and shielding	Alternative for DDQ pads.

Table 5. Contrasting different polar and nonpolar coil configurations

Coil Shape	Structure	Type of Polarization	Flux Creation	Empty Zone
Circular		Unpolarized	A quarter of the pad diameter	40%
DD		Polarized in Parallel	A quarter of the pad diameter over all directions	all around
DDQ	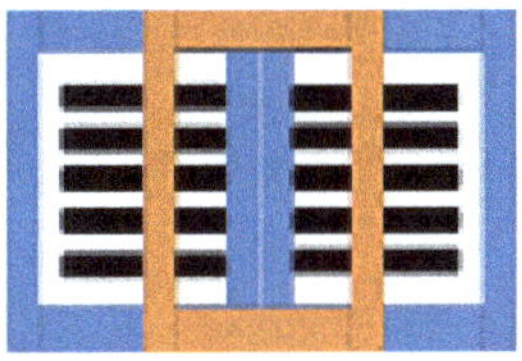	Polarized in Parallel	Half of the pad diameter	34%
Quad D		Polarized, Combined both Parallel and Perpendicular	Two times the sides of the coil	77%
Tripod		Polarized, Parallel and Perpendicular	Two times the sides of the coil	30%

6. CONCLUSION

In this paper, a detailed review of various structures and shapes of the coil for wireless charging has been studied. We conclude compared to the power transfer that is created by the single transmitter coil construction, the hybrid combination of the transmitter will produce a magnetitic field that is significantly stronger than the one produced by the single transmitter coil structure. Also, Structures and coil shapes are compared on various coil parameters and helped us to select a suitable coil structure for a proposed wireless power transfer such as static and dynamic. The following studies were made

1. A detailed study on the types of wireless charging for recent charging of vehicles.
2. A detailed survey on various types of coil structures with pros and cons was made.
3. The impact on efficiency for different coil shapes with respect to different parameters of the coil has been provided with a comparison chart.
4. Assistance for selecting a suitable structure for a proposed wireless power transfer has also been provided.

REFERENCES

Ahmad, A., Alam, M. S., & Chabaan, R. (2018). A comprehensive review of wireless charging technologies for electric vehicles. *IEEE Transactions on Transportation Electrification*, *4*(1), 38–63. DOI: 10.1109/TTE.2017.2771619

Ahmad, A., Khalid, M., Ullah, Z., Ahmad, N., Aljaidi, M., Malik, F. A., & Manzoor, U. (2022). Electric vehicle charging modes, technologies and applications of smart charging. *Energies*, *15*(24), 9471. Advance online publication. DOI: 10.3390/en15249471

Alanazi, F. (2023). Electric vehicles: Benefits, challenges, and potential solutions for widespread adaptation. *Applied Sciences (Basel, Switzerland)*, *13*(10), 6016. Advance online publication. DOI: 10.3390/app13106016

Azad, A. N., Echols, A., Kulyukin, V. A., Zane, R., & Pantic, Z. (2019). Analysis, optimization, and demonstration of a vehicular detection system intended for dynamic wireless charging applications. *IEEE Transactions on Transportation Electrification*, *5*(1), 147–161. DOI: 10.1109/TTE.2018.2870339

Bosshard, R., Mühlethaler, J. W., Kolar, J. W., & Stevanović, I. (2013). Optimized magnetic design for inductive power transfer coils. In *2013 Twenty-Eighth Annual IEEE Applied Power Electronics Conference and Exposition (APEC)* (pp. 1812–1819). IEEE. DOI: 10.1109/APEC.2013.6520541

Bucher, A., & Duerbaum, T. (2012). Analysis and design of a contactless power transmission system based on the extended first harmonic approximation. In *2012 IEEE Industry Applications Society Annual Meeting* (pp. 1–7). IEEE. DOI: 10.1109/IAS.2012.6374114

Budhia, M., Boys, J. T., Covic, G. A., & Huang, C.-Y. (2011). Development of a single-sided flux magnetic coupler for electric vehicle IPT charging systems. *IEEE Transactions on Industrial Electronics*, *60*(1), 318–328. DOI: 10.1109/TIE.2011.2179274

Budhia, M., Covic, G. A., & Boys, J. T. (2011). Design and optimization of circular magnetic structures for lumped inductive power transfer systems. *IEEE Transactions on Power Electronics*, *26*(11), 3096–3108. DOI: 10.1109/TPEL.2011.2143730

Chowdary, K. V., Kumar, K., Behera, R. K., & Banerjee, S. (2020). Overview and analysis of various coil structures for dynamic wireless charging of electric vehicles. In *2020 IEEE International Conference on Power Electronics, Smart Grid and Renewable Energy (PESGRE2020)* (pp. 1–6). IEEE. DOI: 10.1109/PESGRE45664.2020.9070496

Chowdhury, M. S. A., & Liang, X. (2019a). Comparative characteristic analysis of circular and double D power pads for electric vehicle wireless charging systems. In *2019 IEEE Canadian Conference of Electrical and Computer Engineering (CCECE)* (pp. 1–4). IEEE. DOI: 10.1109/CCECE.2019.8861733

Chowdhury, M. S. A., & Liang, X. (2019b). Power transfer efficiency evaluation of different power pads for electric vehicle's wireless charging systems. In *2019 IEEE Canadian Conference of Electrical and Computer Engineering (CCECE)* (pp. 1–4). IEEE. DOI: 10.1109/CCECE.2019.8861757

Cota, K. A., Gray, P. A., Pathmanathan, M., & Lehn, P. W. (2019). An approach for selecting compensation capacitances in resonance-based EV wireless power transfer systems with switched capacitors. *IEEE Transactions on Transportation Electrification*, *5*(4), 1004–1014. DOI: 10.1109/TTE.2019.2927803

Covic, G. A., & Boys, J. T. (2013). Modern trends in inductive power transfer for transportation applications. *IEEE Journal of Emerging and Selected Topics in Power Electronics*, *1*(1), 28–41. DOI: 10.1109/JESTPE.2013.2264473

Covic, G. A., Boys, J. T., Budhia, M., & Huang, C.-Y. (2010). Electric vehicles–personal transportation for the future. *World Electric Vehicle Journal*, *4*(4), 693–704. DOI: 10.3390/wevj4040693

Dai, X., Li, X., Li, Y., & Hu, A. P. (2018). Maximum efficiency tracking for wireless power transfer systems with dynamic coupling coefficient estimation. *IEEE Transactions on Power Electronics*, *33*(6), 5005–5015. DOI: 10.1109/TPEL.2017.2729083

Deng, J., Lu, F., Li, S., Nguyen, T.-D., & Mi, C. (2014). Development of a high efficiency primary side controlled 7kW wireless power charger. In *2014 IEEE International Electric Vehicle Conference (IEVC)* (pp. 1–6). IEEE. DOI: 10.1109/IEVC.2014.7056204

Diep, N. T., Trung, N. K., & Minh, T. T. (2019). Maximum efficiency in the dynamic wireless charging systems of electric vehicles. In 2019 10th International Conference on Power Electronics and ECCE Asia (ICPE 2019—ECCE Asia) (pp. 1–6). IEEE. DOI: 10.23919/ICPE2019-ECCEAsia42246.2019.8797184

Diep, N. T., Trung, N. K., & Minh, T. T. (2020). Wireless power transfer system design for electric vehicle dynamic charging application. *International Journal of Power Electronics and Drive Systems*, *11*(3), 1468–1480. DOI: 10.11591/ijpeds.v11.i3.pp1468-1480

Ehsani, M., Gao, Y., & Emadi, A. (2010). *Modern electric, hybrid electric, and fuel cell vehicles: Fundamentals, theory, and design* (2nd ed.). CRC Press., DOI: 10.1201/9781420054002

Emadi, A., Ehsani, M., & Miller, J. M. (2003). *Vehicular electric power systems: Land, sea, air, and space vehicles*. CRC Press., DOI: 10.1201/9780203913468

Etacheri, V., Marom, R., Elazari, R., Salitra, G., & Aurbach, D. (2011). Challenges in the development of advanced Li-ion batteries: A review. *Energy & Environmental Science, 4*(9), 3243–3262. DOI: 10.1039/c1ee01598b

Feng, Y., & Lu, X. (2021). Construction planning and operation of battery swapping stations for electric vehicles: A literature review. *Energies, 14*(24), 8202. Advance online publication. DOI: 10.3390/en14248202

Gao, Y., Duan, C., Oliveira, A. A., Ginart, A., Farley, K. B., & Tse, Z. T. H. (2017). 3-D coil positioning based on magnetic sensing for wireless EV charging. *IEEE Transactions on Transportation Electrification, 3*(3), 578–588. DOI: 10.1109/TTE.2017.2696787

Hata, K., Imura, T., & Hori, Y. (2016). *Dynamic wireless power transfer system for electric vehicles to simplify ground facilities—power control and efficiency maximization on the secondary side. In 2016 IEEE Applied Power Electronics Conference and Exposition (APEC)*. IEEE., DOI: 10.1109/APEC.2016.7468105

Hossain, A., Darvish, P., Mekhilef, S., Tey, K. S., & Tong, C. W. (2022). A new coil structure of dual transmitters and dual receivers with integrated decoupling coils for increasing power transfer and misalignment tolerance of wireless EV charging system. *IEEE Transactions on Industrial Electronics, 69*(8), 7869–7878. DOI: 10.1109/TIE.2021.3108697

Ibrahim, M., Pichon, L., Bernard, L., Razek, A., Houivet, J., & Cayol, O. (2015). Advanced modeling of a 2-kW series–series resonating inductive charger for real electric vehicle. *IEEE Transactions on Vehicular Technology, 64*(2), 421–430. DOI: 10.1109/TVT.2014.2325614

Jamakani, E., Afjei, E., & Mosallanejad, A. (2019). A novel triple quadrature pad for inductive power transfer systems for electric vehicle charging. In 2019 10th International Power Electronics, Drive Systems and Technologies Conference (PEDSTC) (pp. 618–623). IEEE. DOI: 10.1109/PEDSTC.2019.8697654

Jeong, S. Y., Kwak, H. G., Jang, G. C., Choi, S. Y., & Rim, C. T. (2018). Dual-purpose nonoverlapping coil sets as metal object and vehicle position detections for wireless stationary EV chargers. *IEEE Transactions on Power Electronics, 33*(9), 7387–7397. DOI: 10.1109/TPEL.2017.2765521

Jiwariyavej, V., Imura, T., & Hori, Y. (2015). Coupling coefficients estimation of wireless power transfer system via magnetic resonance coupling using information from either side of the system. *IEEE Journal of Emerging and Selected Topics in Power Electronics, 3*(1), 191–200. DOI: 10.1109/JESTPE.2014.2332056

Kalwar, K. A., Mekhilef, S., Seyedmahmoudian, M., & Horan, B. (2016). Coil design for high misalignment tolerant inductive power transfer system for EV charging. *Energies, 9*(11), 937. Advance online publication. DOI: 10.3390/en9110937

Kan, T., Lu, F., Nguyen, T.-D., Mercier, P. P., & Mi, C. C. (2018). Integrated coil design for EV wireless charging systems using LCC compensation topology. *IEEE Transactions on Power Electronics, 33*(11), 9231–9241. DOI: 10.1109/TPEL.2018.2794448

Kim, S., Covic, G. A., & Boys, J. T. (2016a). Analysis on tripolar pad for inductive power transfer systems. In *2016 IEEE PELS Workshop on Emerging Technologies: Wireless Power Transfer (WoW)* (pp. 15–20). IEEE. DOI: 10.1109/WoW.2016.7772059

Kim, S., Covic, G. A., & Boys, J. T. (2016b). Tripolar pad for inductive power transfer systems for EV charging. *IEEE Transactions on Power Electronics, 32*(7), 5045–5057. DOI: 10.1109/TPEL.2016.2606893

Kobayashi, K., Imura, T., & Hori, Y. (2015). Real-time coupling coefficient estimation and maximum efficiency control on dynamic wireless power transfer for electric vehicles. In *2015 IEEE PELS Workshop on Emerging Technologies: Wireless Power (WoW)* (pp. 1–6). IEEE. DOI: 10.1109/WoW.2015.7132799

Küpper, K., Pels, T., Deiml, M., Angermaier, A., Bürger, T., Weinzerl, A., & Teuschl, G. (2015). Tension 12 V to 800 V efficient powertrain solutions. *Efficient Powertrain Solutions for 12V Up to 800V.*

Laporte, S., Coquery, G., Deniau, V., De Bernardinis, A., & Hautière, N. (2019). Dynamic wireless power transfer charging infrastructure for future EVs: From experimental track to real circulated roads demonstrations. *World Electric Vehicle Journal, 10*(4), 84. Advance online publication. DOI: 10.3390/wevj10040084

Lin, F. Y., Carretero, C., Covic, G. A., & Boys, J. T. (2016). A reduced order model to determine the coupling factor between magnetic pads used in wireless power transfer. *IEEE Transactions on Transportation Electrification*, *3*(2), 321–331. DOI: 10.1109/TTE.2016.2628793

Lin, F. Y., Covic, G. A., & Kesler, M. (2020). Design of a SAE compliant multicoil ground assembly. *IEEE Journal of Emerging and Selected Topics in Industrial Electronics*, *1*(1), 14–25. DOI: 10.1109/JESTIE.2020.2999597

Lin, F. Y., Zaheer, A., Budhia, M., & Covic, G. A. (2014). *Reducing leakage flux in IPT systems by modifying pad ferrite structures. In 2014 IEEE Energy Conversion Congress and Exposition (ECCE)*. IEEE., DOI: 10.1109/ECCE.2014.6953630

Lin, F. Y., Zaheer, A., Budhia, M., & Covic, G. A. (2015). Evaluation of magnetic pad sizes and topologies for electric vehicle charging. *IEEE Transactions on Power Electronics*, *30*(11), 6391–6407. DOI: 10.1109/TPEL.2015.2419592

Liu, C., Jiang, C., & Qiu, C. (2017). Overview of coil designs for wireless charging of electric vehicle. In *2017 IEEE PELS Workshop on Emerging Technologies: Wireless Power Transfer (WoW)* (pp. 1–6). IEEE. DOI: 10.1109/WoW.2017.7959389

Liu, N., & Habetler, T. G. (2013). A study of designing a universal inductive charger for electric vehicles. In IECON 2013—39th Annual Conference of the IEEE Industrial Electronics Society (pp. 4528–4533). IEEE. DOI: 10.1109/IECON.2013.6699865

Madawala, U. K., Neath, M., & Thrimawithana, D. J. (2013). A power–frequency controller for bidirectional inductive power transfer systems. *IEEE Transactions on Industrial Electronics*, *60*(1), 310–317. DOI: 10.1109/TIE.2011.2174537

Mohamed, A. A. S., Berzoy, A., de Almeida, F. G. N., & Mohammed, O. (2017). Modeling and assessment analysis of various compensation topologies in bidirectional IWPT system for EV applications. *IEEE Transactions on Industry Applications*, *53*(5), 4973–4984. DOI: 10.1109/TIA.2017.2700281

Mohamed, A. A. S., Shaier, A. A., Metwally, H., & Selem, S. I. (2020). A comprehensive overview of inductive pad in electric vehicles stationary charging. *Applied Energy*, *262*, 114584. Advance online publication. DOI: 10.1016/j.apenergy.2020.114584

Mojumder, M. R. H., Ahmed Antara, F., Hasanuzzaman, M., Alamri, B., & Alsharef, M. (2022). Electric vehicle-to-grid (V2G) technologies: Impact on the power grid and battery. *Sustainability (Basel)*, *14*(21), 13856. Advance online publication. DOI: 10.3390/su142113856

Mosammam, A. M., Rasekh, N., Mirsalim, M., & Moghani, J. S. (2019). *Comparative analysis of the conventional magnetic structure pads for the wireless power transfer applications. In 2019 10th International Power Electronics, Drive Systems and Technologies Conference (PEDSTC).* IEEE., https://emtrl.aut.ac.ir/publication/conference/

Mou, X., Gladwin, D. T., Zhao, R., Sun, H., & Yang, Z. (2020). Coil design for wireless vehicle-to-vehicle charging systems. *IEEE Access : Practical Innovations, Open Solutions*, *8*, 172723–172733. DOI: 10.1109/ACCESS.2020.3025787

Musavi, F., & Eberle, W. (2014). Overview of wireless power transfer technologies for electric vehicle battery charging. *IET Power Electronics*, *7*(1), 60–66. DOI: 10.1049/iet-pel.2013.0047

Panchal, C., Stegen, S., & Lu, J. (2018). Review of static and dynamic wireless electric vehicle charging system. *Engineering Science and Technology, an International Journal*, *21*(5), 922–937. DOI: 10.1016/j.jestch.2018.06.015

Patil, D., McDonough, M. K., Miller, J. M., Fahimi, B., & Balsara, P. T. (2018). Wireless power transfer for vehicular applications: Overview and challenges. *IEEE Transactions on Transportation Electrification*, *4*(1), 3–37. DOI: 10.1109/TTE.2017.2780627

Petersen, M., & Fuchs, F. W. (2014). Development of a 5 kW inductive power transfer system including control strategy for electric vehicles. In 2014 International Exhibition and Conference for Power Electronics, Intelligent Motion, Renewable Energy and Energy Management (PCIM) (pp. 1–8). IEEE. DOI: 10.1109/JPROC.2013.2244536

Pevere, A., Petrella, R., Mi, C. C., & Zhou, S. (2014). Design of a high efficiency 22 kW wireless power transfer system for EVs fast contactless charging stations. In *2014 IEEE International Electric Vehicle Conference (IEVC)* (pp. 1–7). IEEE. DOI: 10.1109/IEVC.2014.7056221

Ramezani, A., & Narimani, M. (2021). A modular coil design for high-power wireless EV charging. In IECON 2021—47th Annual Conference of the IEEE Industrial Electronics Society (pp. 1–5). IEEE. DOI: 10.1109/IECON48115.2021.9589747

Rosu, S. G., Khalilian, M., Cirimele, V., & Guglielmi, P. (2016). A dynamic wireless charging system for electric vehicles based on DC/AC converters with SiC MOSFET-IGBT switches and resonant gate-drive. In 2016 IEEE 42nd Annual Conference of the IEEE Industrial Electronics Society (IECON) (pp. 4465–4470). IEEE. DOI: 10.1109/IECON.2016.7793809

Shimizu, R., Kaneko, Y., & Abe, S. (2013). A new HE core transmitter of a contactless power transfer system that is compatible with circular core receivers and H-shaped core receivers. In 2013 3rd International Electric Drives Production Conference (EDPC) (pp. 1–7). IEEE. DOI: 10.3390/wevj10030049

Sinha, S., Kumar, A., Regensburger, B., & Afridi, K. K. (2019). A new design approach to mitigating the effect of parasitics in capacitive wireless power transfer systems for electric vehicle charging. *IEEE Transactions on Transportation Electrification*, *5*(4), 1040–1059. DOI: 10.1109/TTE.2019.2931869

Tan, L., Zhao, W., Liu, H., Li, J., & Huang, X. (2020). Design and optimization of ground-side power transmitting coil parameters for EV dynamic wireless charging system. *IEEE Access : Practical Innovations, Open Solutions*, *8*, 74595–74604. DOI: 10.1109/ACCESS.2020.2988622

Tang, Y., Zhu, F., Wang, Y., & Ma, H. (2015). Design and optimizations of solenoid magnetic structure for inductive power transfer in EV applications. In IECON 2015—41st Annual Conference of the IEEE Industrial Electronics Society (pp. 001459–001464). IEEE. DOI: 10.1109/IECON.2015.7392306

Tejeda, C., Carretero, C., Boys, J. T., & Covic, G. A. (2016). Ferrite-less circular pad with controlled flux cancellation for EV wireless charging. *IEEE Transactions on Power Electronics*, *32*(11), 8349–8359. DOI: 10.1109/TPEL.2016.2642192

Tuttle, D. P., & Baldick, R. (2012). The evolution of plug-in electric vehicle-grid interactions. *IEEE Transactions on Smart Grid*, *3*(1), 500–505. DOI: 10.1109/TSG.2011.2168430

Un-Noor, F., Padmanaban, S., Mihet-Popa, L., Mollah, M. N., & Hossain, E. (2017). A comprehensive study of key electric vehicle (EV) components, technologies, challenges, impacts, and future direction of development. *Energies*, *10*(8), 1217. Advance online publication. DOI: 10.3390/en10081217

Villa, J. L., Sallán, J., Llombart, A., & Sanz, J. F. (2009). Design of a high frequency inductively coupled power transfer system for electric vehicle battery charge. *Applied Energy*, *86*(3), 355–363. DOI: 10.1016/j.apenergy.2008.05.009

Yamada, Y., Sasaki, K., Imura, T., & Hori, Y. (2021). Design method of coils for dynamic wireless power transfer considering average transmission power and installation rate. In *2021 IEEE Southern Power Electronics Conference (SPEC)* (pp. 1–8). IEEE. DOI: 10.1109/SPEC52827.2021.9709485

Yang, G., Song, K., Huang, X., Wang, C., Huang, X., Li, J., & Zhang, Y. (2021a). Improved interoperability evaluation method for wireless charging systems based on interface impedance. *IEEE Transactions on Power Electronics, 36*(8), 8588–8592. DOI: 10.1109/TPEL.2021.3053353

Yang, G., Song, K., Sun, Y., Huang, X., Li, J., Guo, Y., Zhang, Y., Zhang, Q., Lu, R., & Zhu, C. (2021b). Interoperability improvement for rectangular pad and DD pad of wireless electric vehicle charging system based on adaptive position adjustment. *IEEE Transactions on Industry Applications, 57*(3), 2613–2624. DOI: 10.1109/TIA.2021.3056639

Yilmaz, M., & Krein, P. T. (2013). Review of battery charger topologies, charging power levels, and infrastructure for plug-in electric and hybrid vehicles. *IEEE Transactions on Power Electronics, 28*(5), 2151–2169. DOI: 10.1109/TPEL.2012.2212917

Yin, J., Lin, D., Parisini, T., & Hui, S. Y. (2016). Front-end monitoring of the mutual inductance and load resistance in a series-series compensated wireless power transfer system. *IEEE Transactions on Power Electronics, 31*(10), 7339–7352. DOI: 10.1109/TPEL.2015.2509962

Zaheer, A., Hao, H., Covic, G. A., & Kacprzak, D. (2014). Investigation of multiple decoupled coil primary pad topologies in lumped IPT systems for interoperable electric vehicle charging. *IEEE Transactions on Power Electronics, 30*(4), 1937–1955. DOI: 10.1109/TPEL.2014.2329693

Zaheer, A., Kacprzak, D., & Covic, G. A. (2012). *A bipolar receiver pad in a lumped IPT system for electric vehicle charging applications. In 2012 IEEE Energy Conversion Congress and Exposition (ECCE).* IEEE., DOI: 10.1109/ECCE.2012.6342811

Zhang, B., Chen, Q., Zhang, L., Chen, J., Xu, L., Ren, X., & Zhang, Z. (2021). Triple-coil-structure-based coil positioning system for wireless EV charger. *IEEE Transactions on Power Electronics, 36*(12), 13515–13525. DOI: 10.1109/TPEL.2021.3085604

Zhang, W., Mo, Q., Yan, P., & Liu, Y. (2021). Simultaneous metal object detection and coil alignment for wireless EV chargers using planar coil array. *IEEE Transactions on Instrumentation and Measurement, 70*, 4507411. Advance online publication. DOI: 10.1109/TIM.2021.3126394

Zhang, W., White, J. C., Abraham, A. M., & Mi, C. C. (2015). Loosely coupled transformer structure and interoperability study for EV wireless charging systems. *IEEE Transactions on Power Electronics, 30*(11), 6356–6367. DOI: 10.1109/TPEL.2015.2433678

Zhang, X., Yuan, Z., Yang, Q., Li, Y., Zhu, J., & Li, Y. (2016). Coil design and efficiency analysis for dynamic wireless charging system for electric vehicles. *IEEE Transactions on Magnetics*, *52*(7), 8700404. Advance online publication. DOI: 10.1109/TMAG.2016.2529682

Chapter 4
The Transformative Potential of Digital Literacy for Redressing Transport Inequality in Emerging Countries

Elena Kiseleva
https://orcid.org/0000-0002-6334-5562
Peter the Great St. Petersburg Polytechnic University, Russia

Andrei Somov
https://orcid.org/0009-0004-2592-9198
Peter the Great St. Petersburg Polytechnic University, Russia

Anna Sedyakina
https://orcid.org/0000-0003-2849-4519
Peter the Great St. Petersburg Polytechnic University, Russia

Lyu Ye
https://orcid.org/0000-0003-2730-7734
Sichuan University of Science and Engineering, China

ABSTRACT

This chapter explores the critical role of digital literacy in smart cities and its impact on economic growth and inequality reduction in emerging countries, focusing on BRICS+ nations. It examines three key areas: smart transport infrastructure, urban mobility, and digital literacy. The study identifies strong correlations between urban mobility indicators and GDP per capita, highlighting digital inclusion as a driver

DOI: 10.4018/979-8-3373-6335-6.ch004

Copyright © 2026, IGI Global Scientific Publishing. Copying or distributing in print or electronic forms without written permission of IGI Global Scientific Publishing is prohibited.

of economic prosperity. Conversely, transport infrastructure shows weaker or mixed effects but significantly correlates with lower inequality. Regression analyses for China, India, and Russia confirm the predominance of digital factors over transport variables in fostering sustainable urban mobility. Achieving near-universal internet access is shown to mitigate urban–rural disparities, especially in regions with underdeveloped transport systems. Government initiatives like "Digital India" and investments in mobile connectivity in the UAE and Iran further emphasize the role of digital transformation in improving accessibility.

INTRODUCTION

Digital literacy is a crucial skill set in the contemporary era as a significant driver of economic development as well as improving the well-being of the world's population. All cities around the world are getting smarter. The smart cities strategy continues to evolve and spread around the world, making the planet more technologically advanced and smarter. There are more and more sustainable urban solutions in agriculture, road's traffic management, reduction of the hydrocarbon footprint of transport, and urban greening. All of these factors enhance the overall sustainability and manufacturability of cities due to smart advanced technologies. Smart technologies have a great advantage and digital literacy is required for its implementation. Increasing digital literacy in emerging countries will contribute to improving for sustainable urban mobility for all citizens without any inequality based on socio-economic status, race or ethnicity. That is why, with the growth of smart urban transportations in new cities, it is important to use the opportunities of digital technologies for increasing the digital literacy of the population in emerging countries. In this context the chapter highlights the important role of digital literacy for smart city and provides valuable insights for emerging countries aimed at transitioning toward environmentally friendly, low-carbon, intelligent and advanced transport.

The chapter includes a study of various aspects of the above: on the one hand, we will study the level of digital literacy and its impact on prosperity and reducing any inequality of urban mobility based on socio-economic status, race or ethnicity. On the other hand, we will consider the possibilities of increasing urban mobility using smart transport, due to the growth of digital literacy of the population in emerging countries.

First of all, the study considers the importance of critical thinking in digital contexts and the emerging field of digital wellbeing are addressed and the components of digital literacy are given. The chapter is investigated the main challenges in achieving digital literacy are diverse barriers such as access to technology, affordability, and cultural differences and others. Specific challenges in mobilizing the transformative potential of digital literacy for BRICS+ countries are considered. The study proves that various digital literacy in emerging countries have significant influence for inclusive growth in BRICS+. For instance, according to Russian Public Opinion Research Center in 2023 the Russia has the level of digital literacy stands at 66% among the adult population and these estimates need to be clarified. The chapter is analyzed the impact of digital literacy on the standard of living of the population in emerging countries, contributing to greater overall household prosperity. By establishing the causal pathways between digital literacy and economic well-being, this study provides research evidence for government to design targeted interventions for reducing digital inequality among the population. Nowadays research community has a dipper understanding that traditional transportation development has significant economic and social benefits, but it has also need to increased pollution emissions and substantial environmental impacts. Obviously, smart transport has all the advantages of traditional transport combined with new smart perspective on reducing carbon emission. Globally, the smart urban mobility market reached $98.8 billion in 2023, with a projected growth to $236.5 billion by 2028 (Grand View Research, n.d.). According to McKinsey report, the implementation of smart city technologies in Moscow's transportation system has reduced average travel time by 10% and decreased the number of road incidents by 23% over the past five years. Similar research based on empirical data from the BRICS's countries is also required. That is why the study examines the variables which are responsible for the logistical development of emerging countries and the impact of smart transport infrastructure on the household prosperity.

First, the integration of these technologies enables the Governments of emerging countries to develop more effective and evidence-based public transport plans for the construction the road transport systems. Second, citizens have access to better information about public transport in real time for trip planning and, thus, can make more effective or inexpensive decisions regarding their daily mobility. According to a Deloitte study, in 2023, 92% of Russians use smartphones, which is 45% more than in 2013. Similar research based on empirical data from the BRICS's countries is also required. The study evaluates whether the use of smart urban mobility depends on the socio-economic characteristics and predicts the level of digital literacy based on statistical data of one of the BRICS's countries. As well as variables such as transport infrastructure, education level, broadband Internet coverage, the number of mobile Internet users and many others. Previously researches are shown that in Russia, more than 85% of users of mobile applications for urban mobility have higher education, confirming the correlation between education level and the use of digital services. According to the Russian Ministry of Digital Development, by the end of 2023, more than 70% of public services in Russian cities were accessible through mobile applications, significantly exceeding the average for BRICS countries, which stands at 45%.[1]We statistically see that the use of mobile technologies has increased significantly over the past 10 years, both in developed and emerging countries, but there are a number of studies which show for the urban mobility is still distributed in favor of those who are more knowledgeable (i.e., tech-savvy). Knowledgeableness is linked to the level of digital literacy. As a result, development of smart transport around the world by using intelligence technology be able to stimulate smart urban mobility of the population and would be able to solve the problem of urban mobility inequality and increase household prosperity in emerging countries. And result of this study, which is shown down, have been proved it.

Digital literacy is a crucial skill set in the contemporary era as a significant driver of economic development as well as improving the well-being of the world's population. All cities around the world are getting smarter. The smart cities strategy continues to evolve and spread around the world, making the planet more technologically advanced and smarter. There are more and more sustainable urban solutions in agriculture, road's traffic management, reduction of the hydrocarbon footprint of transport, and urban greening. All of these factors enhance the overall sustainability and manufacturability of cities due to smart advanced technologies. Smart technologies have a great advantage and digital literacy is required for its implementation. Increasing digital literacy in emerging countries will contribute to improving for sustainable urban mobility for all citizens without any inequality. That is why, with the growth of smart urban transportations in new cities, it is important to use the opportunities of digital technologies for increasing the digital literacy of the population in emerging countries. In this context the chapter highlights the important role of digital literacy for smart city and provides valuable insights for emerging countries aimed at transitioning toward environmentally friendly, low-carbon, intelligent and advanced transport.

LITERATURE REVIEW

The transformative potential of digital literacy for redressing transport inequality in emerging countries has been widely addressed through interdisciplinary perspectives. A smart transportation system architecture that integrates IoT and big data to optimize mobility has been proposed, offering significant promise for developing regions (Jan et al., 2019). A comparative study of megacities in India identified structural urban transport deficiencies exacerbated by socio-economic divides (Reddy & Balachandra, 2012). The influence of digital infrastructures on urban governance and equitable transport access has also been highlighted (Barns et al., 2016). Digital infrastructure has been shown to promote green enterprise transformations, indirectly supporting more sustainable transport models (Guo et al., 2024). Future-focused perspectives on urban mobility and public transport emphasize digital integration as a driver of service inclusivity (Ceder, 2020).

The importance of a sustainable digital ecosystem, where digital servitization enhances infrastructure inclusivity, has been emphasized (Chen et al., 2023). Smart transportation fundamentals with implications for reducing spatial inequalities in mobility have been outlined (Mrityunjaya et al., 2017). Digital infrastructure investment is linked with broader economic and transport benefits, particularly in underserved areas (Greenstein, 2021). Definitional divergences in sustainable urban mobility suggest that stakeholder perceptions shape policy effectiveness (Foltýnová et al., 2020). The generative mechanisms behind digital infrastructure evolution are relevant for long-term transport innovation (Henfridsson & Bygstad, 2013). An evaluation system for smart transport in Chinese cities proposes self-organizing principles crucial for emerging contexts (Yan et al., 2020). Systemic insights into smart transport design within smart city frameworks have also been presented (Jimenez, 2018). A typology of digital infrastructure growth tactics informs adaptive mobility strategies (Koutsikouri et al., 2018). A comprehensive review of smart urban mobility innovations maps their impact on transport equity (Butler et al., 2020). Digital infrastructure construction yields energy savings, reinforcing environmental and social sustainability (Wang & Shao, 2024).

Urban mobility scenarios through 2030 have been modeled, integrating digital innovation trajectories (Miskolczi et al., 2021). Urban mobility indexes are reviewed as tools for quantifying inequality (Costa et al., 2017). A foundational study on urban transport in developing regions highlights systemic constraints (Gakenheimer, 1999). The evaluation of urban form and transport patterns underscores the social costs of exclusionary infrastructure (Camagni et al., 2002). Sustainable mobility is framed as a temporally evolving construct, deeply intertwined with digital capacities (Schipper et al., 2020).

Digital systems in smart cities have been analyzed, positioning "Digital as a Service" as essential for inclusive infrastructure (Serrano, 2018). Performance metrics in smart transport management are linked to user equity (Stefansson & Lumsden, 2009). Key themes around spatial scale in sustainable mobility advocate for digital planning tools (Van Wee & Handy, 2015). Smart transport in global innovation management is examined, connecting it to localized digital competencies (Verma et al., 2020). Digital infrastructure enables collaborative innovation, essential for regional transport cohesion (Tian & Lu, 2023). Demand management strategies in smart transport reinforce the role of data-informed digital literacy (Qin et al., 2022). An architectural overview of smart transport systems pinpoints critical enabling technologies (Xu et al., 2017). Planning models and algorithms for smart transport highlight digital literacy as a prerequisite for effective system use (Karami & Kashef, 2020). The relationship between smart transportation and green development efficiency is questioned, with implications for equitable mobility (Zhao et al., 2023). The digital divide is investigated, with a caution that innovation without literacy can exacerbate inequalities rather than reduce them (Du & Wang, 2024).This body of literature affirms that digital literacy acts as a transformative lever in mitigating transport inequality. It empowers citizens to access, understand, and leverage transport technologies, enhances institutional planning, and ensures that digital infrastructure investment translates into inclusive mobility. The synergy between technological capability and human capital is essential for equitable urban futures in emerging economies.

METHODOLOGY

This section contains the answer to the question "How and why indicators were selected for research?" To assess the emerging field of digital wellbeing under the influence of digital literacy, transport infrastructure, and urban mobility, we developed our methodology as follows.

First of all, we divided all factors affecting digital development into four blocks:

- economic wellbeing of population indicators: reflecting the prosperity and standard of living;
- urban mobility indicators: reflecting development of mobile communications and Internet access;
- smart transport: indicators reflecting the quality of logistics infrastructure;
- digital literacy: indicators reflecting the ability of individuals to effectively communicate using digital technologies.

For each of these blocks, we identified indicators reflecting that block's level of development in the BRICS countries. The next step statistical data for the period from 2000 to 2023 are collected. After that addressed gaps in the data using extrapolation, and thus obtained the research dataset. The next step involved constructing correlation relationships between the parameters to examine the influence of urban mobility, smart transport, and digital literacy on assessing the emerging field of digital wellbeing for each BRICS+ countries over randomly selected intervals. We then analyzed the resulting data and summarized the final findings for the dependent and independent variables. The concluding stage of the study entailed building regression relationships for each BRICS country. Below, the chapter is discussed each stage of the research in more detail.

Table 1. Indicators for assessing the level of smart transport development

Indicator	Explanation	In the BRICS context
Economic wellbeing of population		
GDP per capita	A basic economic indicator. Comparing the dynamics of this indicator makes it possible to understand which countries are making more significant progress in economic development.	BRICS countries vary greatly in terms of GDP per capita: China and Russia have higher values, while India and South Africa are lower. Brazil is somewhere in between.
Gini index (GI)	Shows the degree of inequality in the distribution of income (or consumption) in society. A decline in this coefficient indicates an increase in the population's well-being and better social stability. .	Some of the BRICS countries have some of the highest levels of inequality in the world (for example, South Africa and Brazil). Tracking the dynamics of the Gini index is important for understanding to what extent economic growth "trickles down" to all segments of the population in the BRICS
Human Capital Index (HCI)	An integral indicator that includes digital literacy and the health of the population. A high human capital index means that people have access to quality healthcare and education, as well as the skills needed for a modern digital and technological economy.	China is actively investing in education and science, achieving high results. Russia maintains strong educational traditions but faces the challenge of "brain drain" and uneven distribution of quality healthcare across its territory. India is making major investments in digital skills and engineering education, yet there remains a gap in the quality of available public services. Brazil and South Africa are also actively working to improve the quality of secondary and higher education but face social and infrastructural issues.
Urban mobility		
Mobile cellular subscriptions (per 100 people)	Direct indicator of the penetration of mobile communications. Reflects the availability needed for the development of mobile services: the higher numbers of subscriber accounts, the greater the potential for expanding mobile services(from mobile banking to mobile smart transport).	The BRICS countries include large territories (Russia, Brazil) and highly populous nations (India, China). This creates very uneven conditions for access to mobile communications.
Mobile broadband subscriptions (per 100 people)	A basic metric for analyzing the population's ability to fully use smartphones and mobile online services via broadband (3G/4G/5G) connections. A tool of digital development: high density of mobile broadband access indicates a good infrastructure, which encourages the emergence of innovative mobile services.	In some BRICS countries (India, South Africa), residents still face problems with broadband availability in remote areas. China, Russia, Brazil, and the UAE lead in these indicators. Ethiopia, India, and South Africa lag significantly behind.

continued on following page

Table 1. Continued

Indicator	Explanation	In the BRICS context
Population covered by a mobile network (% of population)	A parameter of geographic coverage: shows which part of the population is physically located within reach of a mobile network. Critically important for analyzing the level of development of digital mobile services. It can also indicate digital inequality.	Even if BRICS countries have a large number of subscribers, there may still be "white spots" without high-speed Internet. This is especially typical of rural, mountainous, or underdeveloped regions (some parts of India, remote areas of Siberia and the Far East in Russia, Brazil's Northeast, etc.).
Individuals using the Internet (% of population)	A basic indicator: the growth of the share of Internet users reflects the increase in "smart" mobile devices and "smart" city wired infrastructure. Access to the Internet is critically important for integration into global digital processes.	In China, mobile payment systems (WeChat Pay, Alipay) have become one of the main drivers of e-commerce. In India, the Digital India program is actively promoting UPI through mobile applications. Egypt and Ethiopia have low levels of financial service adoption, while Russia, Iran, China, and Brazil are among the leaders.
Mobile money accounts (%, age 15+)	This indicator characterizes the transition to being included in the use of mobile services. If a country has a high share of "mobile money accounts," it means the population actively uses mobile services for daily financial transactions (bill payments, purchases, money transfers). It indicates a stimulus (impetus) for further growth of mobile services.	In China, mobile payment systems (WeChat Pay, Alipay) have become one of the main drivers of e-commerce. In India, the Digital India program is actively promoting UPI through mobile applications. Egypt and Ethiopia have low levels of financial service adoption, while Russia, Iran, China, and Brazil are among the leaders.
Smart transport infrastructure		
Logistics Performance Index: Infrastructure	A comprehensive assessment of the quality of logistics infrastructure. The LPI takes into account the quality of roads, railways, ports, airports, and logistics services, including the presence of digital solutions (tracking systems, electronic customs, etc.).	BRICS countries significantly differ in the level of infrastructure and the adoption of digital tools: China, the UAE, Brazil, and Russia have high indicators in major cities but there are issues in remote regions; in India and Ethiopia, South Africa, the logistics infrastructure can lag behind significantly in rural areas.
Railways, passengers carried	An indicator of integration into smart transport systems. It shows how countries are developing high-speed rail lines, automated train movement control systems, e-tickets, etc.	China has one of the largest and most technologically advanced high-speed rail networks. Russia has extensive networks with various degrees of modernization. India, Brazil, and South Africa are actively developing "smart" passenger travel as part of individual projects.
Air transport, passengers carried	Reflects the level of development and innovation of the transport network through the use of "smart" technologies in airports. It is becoming one of the drivers for "smart city" and "smart infrastructure."	Vast distances (Russia, Brazil, China) and a growing middle class (India, China) lead to an increase in the number of air passengers and to higher levels of technological investment in the sector.

continued on following page

Table 1. Continued

Indicator	Explanation	In the BRICS context
Digital literacy		
School enrollment, primary and secondary(% gross)	he level of parameter indicates the accessibility and quality of basic education. It lays the groundwork for digital literacy.	China and Russia have long had a high level of this indicator, whereas India and South Africa show significant variation in education quality across different regions.
Educational attainment, at least Bachelor's or equivalent, population 25+, total (%) (cumulative)	The proportion of the population with higher education reflects the overall level of human capital in a country and reflects in turn increases the demand for digital skills	China has recently made a radical push to increase the share of specialists with technically innovative education. Russia faces the challenge of "brain drain". Brazil is actively developing its private university system, while South Africa and Ethiopia have lower levels of digital literacy.
Literacy rate, youth total (% of people ages 15-24)	The youth literacy rate demonstrates young people's ability to absorb and utilize digital skills.	A high level of literacy among young people in China, Russia, and Brazil ensures readiness for adopting digital solutions. India and Egypt have shown significant growth in young people's digital literacy, but they still lag behind
ICT goods exports and imports (% of total goods exports)	The share of ICT goods in trade indirectly characterizes the technological advancement of the economy and its participation in global ICT value chains.	China is a world leader in ICT goods exports. The raw materials and traditional export commodities in Russia, Brazil, and South Africa are dominated. The wide variation in this indicator reflects each country's prospects for technology exchange and R&D investments, which significantly influences digital literacy level.

To evaluate the environment of digital wellbeing in BRICS+ countries, it is applied three basic (primary) indicators of economic wellbeing of population: GDP per capita, Gini index, Human Capital Index. Although an increase in GDP per capita is usually associated with improvements in the standard of living, in the context of our study it primarily reflects the capacity for investment in BRICS countries—i.e., the possibility of investing in technological development. Since high per capita GDP by itself does not guarantee prosperity for the entire population (it does not reflect the huge income gap between rich and poor), the study is supplemented this block with the Gini coefficient and Human Capital Index. Tracking Gini index dynamics is essential for understanding whether economic growth "reaches" all segments of society. The Human Capital Index helps us see whether the population has access to digital healthcare, online education, and other digital services, which indicates the development of the skills and competencies needed for a modern technological economy.

In the urban mobility block is assessed the population's capacity to use smart transport through the use of mobile communication and internet availability. Number of SIM cards reflects the basic availability of digital mobile services to the population. However, SIM cards by themselves do not guarantee Internet access. Therefore, we add another indicator to track the availability of broadband (3G/4G/5G) connections. High density of mobile broadband access indicates a robust infrastructure that fosters the emergence of innovative services (mobile banking, e-government, telemedicine, etc.). All of innovative services are necessary for introducing smart transport infrastructure. This block of indicators shows the accessibility of digital development, the potential to achieve inclusivity in using various mobile apps, and the effectiveness of government programs encouraging mobile operators to expand networks not only in megacities but also in rural areas. It also reflects the convenience and accessibility of mobile applications in BRICS countries. Authors believe this block effectively distinguishes BRICS+ countries by the level of the population's digital engagement in using mobile services for everyday needs.

In the smart transport infrastructure block is evaluated the availability of digital development in smart transport through the presence of advanced technologies in logistics, urban road networks, and air and rail passenger transport. We use the Logistics Performance Index (LPI) because a high value indicates the presence of advanced technologies for managing transportation flows (electronic documentation, monitoring systems, etc.) in BRICS+ countries, aligning with the concept of smart transport and smart city. By examining passenger traffic on railways and air transport, we assess the level of smart transport infrastructure development in BRICS+ cities, since airport logistics often demands cutting-edge technical solutions—from air traffic control systems to digital passenger services (electronic tickets, smart check-in apps, automation, biometric controls, IoT solutions for baggage, etc.)—as well as green rail transport.

In the digital literacy block, we examine how BRICS+ digital literacy indicators influence the economic wellbeing of these countries' populations. Our objective is to verify the hypothesis that broad coverage of primary and secondary education cultivates the fundamental knowledge required for integration into the economy and public life. We further propose that a larger share of the adult population holding higher education credentials signals a nation's aptitude for generating innovation and mastering complex technological processes, a claim we plan to test empirically. High digital literacy rates among young people are deemed essential for creating a flexible workforce that can readily adopt digital tools and participate in the global knowledge economy. China and Russia have consistently maintained near-universal primary and secondary school enrollment, whereas India has achieved substantial progress through large-scale governmental programs, although the quality of education may vary considerably across different districts, particularly in rural regions. In South Africa, the proportion of individuals who have attained higher education remains comparatively low relative to other BRICS+ nations, yet efforts are underway to expand the university sector to address historical imbalances. Recent data for China, Russia, and Brazil suggest that youth literacy rates approach 95%, indicating a strong capacity to absorb modern technologies, pursue further education, and ultimately facilitate GDP growth and enhance overall welfare. China's high proportion of ICT goods in its export (and import) structure points to the country's advanced technological standing and underscores gaps in digital skills development and IT-sector expansion among other BRICS members.

RESULTS

This section is examined the pivotal role of digital urban mobility in advancing smart city and smart transport concepts by analyzing how the selected indicators from three thematic blocks influence overall prosperity and economic wellbeing of population. We identify those indicators that most strongly enable emerging countries (most of the BRICS+) to harness smart transport benefits for their populations and to reduce any inequalities in urban digital mobility.

Correlations

The next step of research is the calculation of correlations between indicators from different blocks. To do this, four correlation matrixуы have constructed (Table 3,4,5). The results are discussed below.

Human Capital Index correlations (Table 3).

Key interrelationships among the independent variables.

- railways, passengers carried vs. air transport, passengers carried: a correlation of 0.94 indicates that countries with robust passenger railway operations often have a substantial volume of air travel (possibly china leading in both).
- logistics performance index vs. mobile cellular subscriptions: a correlation of 0.53 suggests that logistics and mobile connectivity develop in parallel in urbanized nations.
- individuals using the internet vs. educational attainment: a correlation of 0.7663 reflects a classic pattern: the higher the proportion of adults with university degrees, the more prevalent internet use becomes.
- fixed broadband subscriptions vs. account ownership at a financial institution: a correlation of 0.76 indicates that countries with a more extensive fixed broadband infrastructure also display a higher degree of banking coverage (financial inclusion), implying that digital literacy ultimately affects the ability to leverage smart transport and smart urban infrastructure.

Table 2. Correlation matrix of Human Capital Index

	HCI	Rail-ways	LPI	AT	MSC	FBS	IUI	MMA	SE	ICT exp g	ICT exp s	EA
Human Capital Index (HCI)	1.000											
Railways. passengers carried (million passenger-km). Railways	0.230	1.000										
Logistics performance index. LPI	-0.593	0.369	1.000									
Air transport. passengers carried. AT	0.274	0.943	0.384	1.000								
Mobile cellular subscriptions (per 100 people). MCS	-0.850	-0.008	0.529	0.027	1.000							
Fixed broadband subscriptions (per 100 people). FBS	-0.345	-0.073	0.443	0.058	0.429	1.000						
Individuals using the Internet (% of population). IUI	-0.673	0.217	0.779	0.314	0.709	0.206	1.000					
Account ownership at a financial institution or with a mobile-money-service provider (% of population ages 15+). MMA	-0.099	0.278	0.333	0.309	0.358	0.761	0.040	1.000				
School enrollment. tertiary (% gross). SE	-0.280	0.017	0.171	0.143	0.218	-0.101	0.549	-0.360	1.000			
ICT goods exports (% of total goods exports). ICT exp g	-0.072	-0.241	0.046	-0.105	0.011	0.400	0.099	0.123	0.054	1.000		
ICT service exports (% of service exports. BoP). ICT exp s	0.277	-0.148	-0.305	-0.153	-0.202	0.099	-0.367	0.452	-0.077	0.040	1.000	
Educational attainment. at least Bachelor's or equivalent. population 25+. total (%) (cumulative). EA	-0.387	0.329	0.490	0.378	0.491	-0.255	0.766	-0.123	0.582	-0.091	-0.098	1.000

Table 3. Correlation matrix of GDP per capita

	GDPpc	Rail-ways	LPI	AT	MSC	FBS	IUI	MMA	SE	ICT exp g	ICT exp s	EA
GDP per capita. current US$. (GDPpc)	1.000											
Railways. passengers carried (million passenger-km). Railways	-0.017	1.000										
Logistics performance index. LPI	0.613	0.369	1.000									
Air transport. passengers carried. AT	0.106	0.943	0.384	1.000								
Mobile cellular subscriptions (per 100 people). MCS	0.740	-0.008	0.529	0.027	1.000							
Fixed broadband subscriptions (per 100 people). FBS	0.318	-0.073	0.443	0.058	0.429	1.000						
Individuals using the Internet (% of population). IUI	0.772	0.217	0.779	0.314	0.709	0.206	1.000					
Account ownership at a financial institution or with a mobile-money-service provider (% of population ages 15+). MMA	0.302	0.278	0.333	0.309	0.358	0.761	0.040	1.000				
School enrollment. tertiary (% gross). SE	0.062	0.017	0.171	0.143	0.218	-0.101	0.549	-0.360	1.000			
ICT goods exports (% of total goods exports). ICT exp g	0.185	-0.241	0.046	-0.105	0.011	0.400	0.099	0.123	0.054	1.000		
ICT service exports (% of service exports. BoP). ICT exp s	-0.141	-0.148	-0.305	-0.153	-0.202	0.099	-0.367	0.452	-0.077	0.040	1.000	
Educational attainment. at least Bachelor's or equivalent. population 25+. total (%) (cumulative). EA	0.598	0.329	0.490	0.378	0.491	-0.255	0.766	-0.123	0.582	-0.091	-0.098	1.000

Table 4 Correlation matrix of GDP per capita growth

	GDP pcg	Rail-ways	LPI	AT	MSC	FBS	IUI	MMA	SE	ICT exp g	ICT exp s	EA
GDP per capita growth (annual %). GDP pcg	1.000											
Railways. passengers carried (million passenger-km). Railways	-0.334	1.000										
Logistics performance index. LPI	-0.807	0.369	1.000									
Air transport. passengers carried. AT	-0.347	0.943	0.384	1.000								
Mobile cellular subscriptions (per 100 people). MCS	-0.729	-0.008	0.529	0.027	1.000							
Fixed broadband subscriptions (per 100 people). FBS	-0.251	-0.073	0.443	0.058	0.429	1.000						
Individuals using the Internet (% of population). IUI	-0.837	0.217	0.779	0.314	0.709	0.206	1.000					
Account ownership at a financial institution or with a mobile-money-service provider (% of population ages 15+). MMA	-0.227	0.278	0.333	0.309	0.358	0.761	0.040	1.000				
School enrollment. tertiary (% gross). SE	-0.239	0.017	0.171	0.143	0.218	-0.101	0.549	-0.360	1.000			
ICT goods exports (% of total goods exports). ICT exp g	0.345	-0.241	0.046	-0.105	0.011	0.400	0.099	0.123	0.054	1.000		
ICT service exports (% of service exports. BoP). ICT exp s	0.373	-0.148	-0.305	-0.153	-0.202	0.099	-0.367	0.452	-0.077	0.040	1.000	
Educational attainment. at least Bachelor's or equivalent. population 25+. total (%) (cumulative). EA	-0.576	0.329	0.490	0.378	0.491	-0.255	0.766	-0.123	0.582	-0.091	-0.098	1.00

Table 5 Correlation matrix of Gini index

	GI	Rail-ways	LPI	AT	MSC	FBS	IUI	MMA	SE	ICT exp g	ICT exp s	EA
Gini index. GI	1.000											
Railways. passengers carried (million passenger-km). Railways	0.756	1.000										
Logistics performance index. LPI	0.134	0.369	1.000									
Air transport. passengers carried. AT	0.582	0.943	0.384	1.000								
Mobile cellular subscriptions (per 100 people). MCS	-0.010	-0.008	0.529	0.027	1.000							
Fixed broadband subscriptions (per 100 people). FBS	-0.062	-0.073	0.443	0.058	0.429	1.000						
Individuals using the Internet (% of population). IUI	0.036	0.217	0.779	0.314	0.709	0.206	1.000					
Account ownership at a financial institution or with a mobile-money-service provider (% of population ages 15+). MMA	0.213	0.278	0.333	0.309	0.358	0.761	0.040	1.000				
School enrollment. tertiary (% gross). SE	-0.054	0.017	0.171	0.143	0.218	-0.101	0.549	-0.360	1.000			
ICT goods exports (% of total goods exports). ICT exp g	-0.143	-0.241	0.046	-0.105	0.011	0.400	0.099	0.123	0.054	1.000		
ICT service exports (% of service exports. BoP). ICT exp s	-0.099	-0.148	-0.305	-0.153	-0.202	0.099	-0.367	0.452	-0.077	0.040	1.000	
Educational attainment. at least Bachelor's or equivalent. population 25+. total (%) (cumulative). EA	-0.017	0.329	0.490	0.378	0.491	-0.255	0.766	-0.123	0.582	-0.091	-0.098	1.00

- school enrollment, tertiary vs. educational attainment: a correlation of 0.58 is logical in that countries with a larger student population will gradually raise the proportion of individuals holding higher education credentials.

Figure 1. Dependency Diagram of Smart Transport Development Factors in BRICS+ Countries

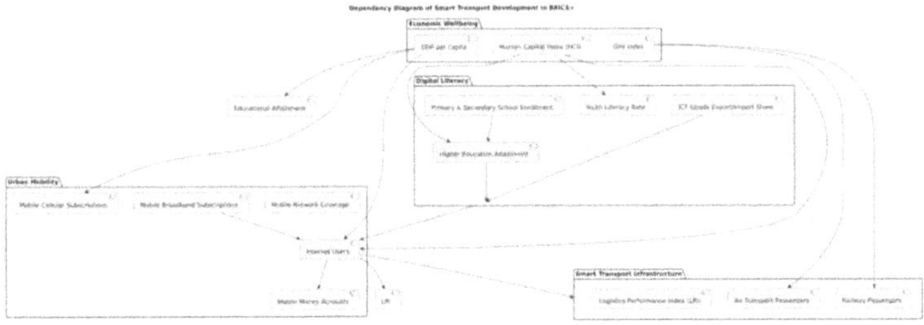

Figure 2. Correlation-Based Relationship Diagram of Digital Wellbeing and Smart Transport in BRICS+

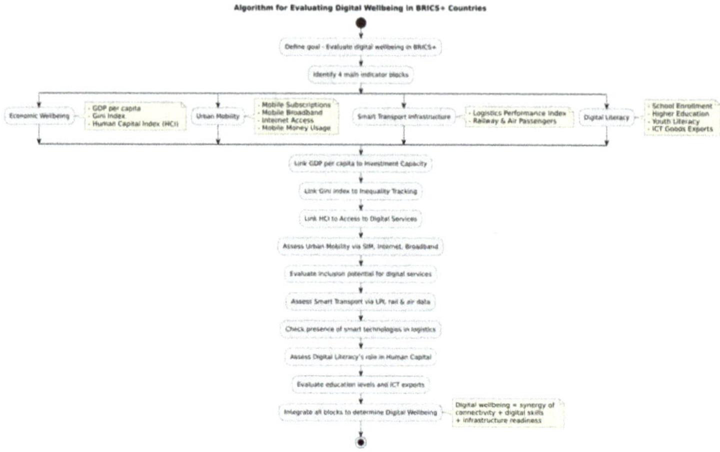

Figure 3. Algorithm for Regression Analysis of Digital, Mobility, and Infrastructure Factors on GDP per Capita in BRICS+

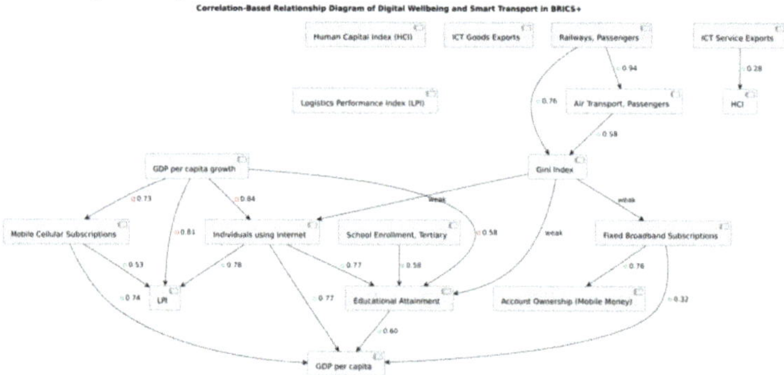

Although Table 2 contains certain anomalies, it also presents several plausible positive associations, potentially reflecting each state's unique developmental path. Crucially, transportation infrastructure (railways, aviation) and IT service exports demonstrate moderate positive relationships with human capital. Meanwhile, mobile and internet indicators in the BRICS context may behave "non-classically," mirroring swift growth in fundamental connectivity in countries with lower HCI (India, South Africa). Educational and digital metrics in the digital literacy block exhibit strong mutual correlations (e.g., higher education associated with more frequent internet use), aligning with standard views on the synergy between education and ICT.

One of the basic economic indicator and it impact on the household prosperity of emerging countries is presented below (Table 3 and Table 4: GDP per capita and GDP per capita growth).

The correlations suggest a pronounced influence of the logistics index (0.61), mobile cellular subscriptions per 100 people (0.74), individuals using the Internet (0.72), and educational attainment (0.6) on GDP per capita. Nations with higher logistics index values tend to exhibit higher per capita GDP, consistent with the hypothesis that strong infrastructure and developed logistics foster economic growth and competitiveness. The robust positive association with mobile internet usage suggests that in BRICS countries with higher per capita GDP, mobile communications are not only extensive in number but also more advanced in quality, potentially reflecting greater purchasing power—supporting the notion that reliable connectivity underpins the expansion of smart transport. A moderate positive correlation with educational attainment indicates that an increase in the proportion of university graduates can enhance a nation's GDP. For instance, Russia and China show comparatively higher shares of tertiary graduates and consequently register higher GDP per capita levels than India or South Africa.

Select correlations among independent variables:

- air transport and Railways, passengers carried: 0.94, implying that countries with well-developed railway passenger traffic (e.g., China) typically also maintain a significant aviation sector.
- individuals using the Internet and Logistics performance index: +0.78, demonstrating that sophisticated infrastructure and logistics indirectly promote internet adoption (urban development, transit networks, commerce, etc.).
- fixed broadband and Account ownership: +0.76, indicating that broader broadband coverage often parallels a more advanced banking sector and greater financial inclusion—logically due to digital channels facilitating banking transactions.

Using two GDP measures confirms that correlation patterns remain valid, with minor coefficient variations. The crucial finding is that GDP per capita strongly aligns with "digital maturity" indicators—robust logistics, widespread internet penetration, extensive mobile subscriptions, and higher educational attainment—while GDP per capita growth is often inversely related, reflecting those countries with established infrastructure and human capital grew at slower relative rates during the observed period. Notably, in the BRICS group, a high GDP per capita most closely correlates with internet penetration, active mobile communication, advanced logistics, and an elevated educational profile. According to the data, these elements are strong markers for deeper digital wellbeing and for developing predictive regression models.

Gini Index (Table 5).

The correlation analysis underscores strong connections of railways (passengers carried, correlation is 0.76) and air transport (passengers carried, correlation is 0.58) with the Gini index. By contrast, there is almost no correlation between Gini and other variables such as Internet penetration (individuals using the Internet), fixed broadband, ICT export shares, or the proportion of higher education. Because the Gini index measures the degree of income and consumption inequality, it likewise suggests digital inequality. This study confirms that countries with large railway passenger volumes (China, Russia) display lower digital inequality (i.e., a lower Gini). Building upon these findings, one can conclude that Brazil, Egypt, Ethiopia, India, and South Africa have higher levels of digital inequality among BRICS+. In this sample, "the higher the railway passenger volume, the lower the digital inequality." However, overall transport accessibility, especially air travel, is also critical, and here we observe an opposite trend: countries with large air travel operations (Iran, India, Indonesia, UAE) exhibit higher levels of digital inequality in remote regions, partly reflecting that rail is generally cheaper than air travel, and countries with highly uneven incomes but a large airline market can maintain elevated income disparity, thus affecting digital inequality.

Furthermore, Egypt, Ethiopia, and South Africa report lower digitalization levels even in major urban centers relative to other BRICS countries. In this dataset, the Gini index correlates most strongly (positively) with rail and air passenger volume, implying that countries with expansive passenger transport infrastructures (India, Brazil, China) may face more severe digital inequality challenges in peripheral areas. Notable associations with urban mobility and digital literacy indicators are scant—most show weak or minimal correlations.

Regressions

The next stage of the study focuses on assessing the impact of digital literacy, developed transport infrastructure, and urban mobility on prosperity, economic wellbeing, and the reduction of digital inequality across various BRICS+ countries. A regression analysis is applied to conduct this assessment. For each BRICS+ country, seven of the most influential variables were selected within three analytical blocks, allowing for the identification of both common and country-specific trends over the period from 2010 to 2023. For some BRICS countries, we were unable to obtain statistically significant models (Table 6).

Table 6. Evaluation of the regression model quality

BRICS+	Adjusted R^2	F-statistics (F-test)	t-statistics (t-test), statistical significance, p-value< 0.05						
			X_1	X_2	X_3	X_4	X_5	X_6	X_7
Brazil	0.68	F = 1.82; p-value 0.241 > 0.05 No	failed the statistical significance test						
China	0.99	F = 56.7; p-value 0.000046 < 0.0001 Yes	0.585 No	0.46 Borderline	0.303 No	0.248 No	0.469 No	0.601 No	0.253 No
India	0.99	F = 61.8; p-value 0.000036 < 0.0001 Yes	0.0017 Yes	0.172 No	0.587 No	0.0006 Yes	0.270 No	0.037 Yes	0.296 No
Ethiopia	0.83	F = 7.94; p-value 0.0576> 0.05 Borderline	failed the statistical significance test						
Russia	0.86	F = 5.54; p-value 0.0266 < 0.05 Yes	0.0066 Yes	0.0531 Borderline	0.0768 No	0.0051 Yes	0.0038 Yes	0.069 No	0.0138 Yes
United Arab Emirates	0.99	F = 367.42; p-value 0.0000001 < 0.0001 Yes	0.23 No	0.0077 Yes	0.014 Yes	0.016 Yes	0.159 No	0.019 Yes	-
Egypt, Arab Rep.	0.99	F = 393.32; p-value 0.0000015 < 0.01 Yes	0.69 No	0.0041 Yes	0.018 Yes	0.015 Yes	0.016 Yes	0.054 Borderline	0.012 Yes
Indonesia	0.99	F = 1079.54; p-value 0.00000012 < 0.0001 Yes	0.29 No	0.77 No	0.46 No	0.16 No	0.99 No	0.0035 Yes	0.038 Yes
South Africa	0.99	F = 131.83, p-value 0.0000229 < 0.0001 Yes	0.58 No	0.32 No	0.90 No	0.31 No	0.79 No	0.067 Borderline	0.093 Borderline
Iran, Islamic Rep.	0.94	F = 15.93; 0.0017 < 0.01 Yes	0.0156 Yes	0.6698 No	0.0926 Borderline	0.0246 Yes	0.0595 Borderline	0.057 Yes	0.8843 No

In other words, the current versions of the models do not permit reliable conclusions about the influence of individual factors on national GDP, since all regression coefficients in the model are statistically insignificant. This applies to Ethiopia and Brazil. Nonetheless, for examples such as China, India, and Russia—key BRICS countries—we were able to draw interesting conclusions. The detailed results of the regression equations for each BRICS+ country are presented below.

China

The resulting linear regression equation for China is as follows:

$$Y=1.09X_1+0.0009X_2+7.3X_3+127.6X_4+78.3X_5+44.9X_6-393.18X_7+31269$$

Symbols: Y - GDP per capita, current USD; X_1 - air transport, passengers carried; X_2 - railways, passengers carried (million passenger-km); X_3 - mobile cellular subscriptions, per 100 people; X_4 - individuals using the internet, % of population; X_5 - fixed broadband subscriptions, per 100 people; X_6 - school enrollment, tertiary, % gross; X_7 - school enrollment, secondary, % gross.

This equation was tested for statistical significance and demonstrates highly reliable results: $R^2=0.98$, and the F-test, along with the t-statistics for certain coefficients, confirms model robustness. The mean approximation error is 2.9%.

The equation suggests that in the case of China—as a BRICS country—the greatest influence on the economic wellbeing of populations stems from urban mobility and digital literacy variables.

Most notably, access to network and broadband internet appears to have the largest impact. The "smart transport" technologies implemented in airport and railway logistics appear less significant for China and exert a smaller effect on prosperity and reducing any inequality of urban mobility. The sizable positive coefficients for Individuals using the Internet also reflect the strong importance of digitalization in boosting China's per capita GDP at present. A negative value for School enrollment may arise from specific structural effects during the reviewed period. Given that China witnessed a large-scale expansion of the mobile communications sector (transitioning from 3G to 5G) in that time, the model indicates that this shift contributed more significantly to the advancement of digital skills and overall prosperity. Overall, the study's resulting equation illustrates how transport metrics (aviation, railways), communication factors (mobile, internet, broadband), and educational parameters (secondary and tertiary enrollment) collectively correlate with changes in China's per capita GDP from 2010 to 2023. Positive coefficients indicate a direct contribution to income growth, whereas negative coefficients may reflect structural idiosyncrasies or temporary "distortions" characteristic of a specific period.

Russia

The developed the regression equation for Russia is shown below. This equation has likewise passed significance testing and demonstrates strong reliability: $R^2=0.87$. Both the F-test and t-statistics attest to its reliability, and the mean approximation error is 6.2%.

$$Y = -0.0003X_1 + 0.012X_2 + 92.3X_3 - 1443.7X_4 + 2660.2X_5 + 2542.1X_6 - 1805.88X_7 + 342116$$

According to the equation, fixed broadband subscriptions, individuals using the internet, and school enrollment hold the most influence on the economic wellbeing of Russia's population. This finding suggests that urban mobility and digital literacy are major factors underpinning the development of a smart, digital urban transport infrastructure. The key drivers of population prosperity include broadband internet, higher education, and accessible mobile communications. The quality of the model, however, is affected by paradoxical coefficients (the negative impact of X_4 and X_7) and an overly large constant term (+342,116)—possibly indicating biases due to unaccounted factors. On the whole, the model is valuable yet would benefit from refinement.

India

The obtained the regression equation for India is:

$$Y=-0.0008X_1-0.0007X_2+3.9X_3-27.93X_4-181.9X_5+8.94X_6-17.34X_7-1234,$$

This equation has undergone significance testing and similarly shows strong reliability: $R^2=0.98$. F- statistics and t-statistics confirm the model's reliability, with a mean approximation error of 1.8%.

The equation reveals that for India—as a BRICS country—individuals using the Internet and school enrollment exert the greatest influence on economic wellbeing, whereas logistics variables have a much smaller impact. It therefore appears that access to mobile internet and digital literacy hold the most decisive roles, aligning with outcomes from other BRICS countries. Strong negative coefficients may signify that in the years witnessing the fastest growth of these metrics, India's GDP per capita grew more modestly, possibly due to delayed investment effects or macroeconomic factors (global trends, currency volatility, etc.). Yet the high predictive power of this model—evidenced by solid explanatory capacity, three statistically significant predictors, and a small standard error—should be noted. However, the presence of non-significant variables and negative coefficients warrants further analysis. Regarding India's results, numerous events influenced the nation's socio-economic climate from 2010 to 2023, including "Digital India," the 2016 demonetization, the COVID-19 pandemic, and global oil-price fluctuations. India also posted significant growth in mobile communications and massive expansions in internet penetration during this period, while still facing infrastructure challenges (rail, aviation) and educational disparities.

United Arab Emirates

The obtained the regression equation for United Arab Emirates is:

$$Y = 0.0289X_1 - 408.1X_2 + 848.88X_3 - 705.93X_4 - 704.82X_5 + 1469.65X_6 + 240855$$

This equation has demonstrated an exceptionally high level of reliability, with $R^2 = 0.997$, indicating that the model explains nearly all the variance in GDP per capita. The F-statistic (367.4) and a significance level of 0.000000199 confirm the robustness of the model, while the normalized R^2 (0.85) and a standard error of 3572.95 provide additional validation. The mean approximation error is acceptably low, indicating a good fit.

According to the model, school enrollment in tertiary education (X_6) and mobile cellular subscriptions (X_3) are the strongest positive contributors to economic well-being in the UAE. These suggest that investment in higher education and widespread access to mobile technology are crucial factors driving economic performance. The air transport factor (X_1) has a modest positive impact, likely reflecting the country's global connectivity. Notably, the negative coefficients for rail passenger traffic (X_2), internet usage (X_4), and fixed broadband subscriptions (X_5) are somewhat counterintuitive. These may reflect unique national characteristics, such as limited reliance on rail infrastructure or the substitution effects between mobile and fixed internet usage in a highly digitized economy. The large constant term (+240,855) may indicate structural factors such as oil wealth, foreign investments, or tax advantages not captured by the variables.

Overall, the model highlights the importance of tertiary education, mobile infrastructure, and global connectivity as core pillars of the UAE's economic prosperity. Nevertheless, further refinement may be necessary to fully account for structural specifics of the Emirati economy.

Egypt, Arab Republic

The developed a regression equation for Egypt is:

$$Y = -0.00168X_1 - 0.2226X_2 + 55.69X_3 + 90.96X_4 - 594.53X_5 - 68.71X_6 + 159.95X_7 + 104036$$

This model has shown a very high explanatory power, with $R^2 = 0.998$, indicating an almost perfect fit to the data. The F-statistic (393.32) and the very low significance value (0.0000015) confirm the model's validity. The normalized R^2 (0.83) and standard error (220.24) support the robustness of the findings.

The most impactful positive contributors to GDP per capita in Egypt, according to the regression, are school enrollment in tertiary education (X_7), internet usage (X_4), and mobile cellular subscriptions (X_3). These factors highlight the country's growing reliance on education and digital inclusion as key drivers of economic well-being. At the same time, railways (X_2) and fixed broadband subscriptions (X_5) exert a negative influence on GDP per capita in this model. This might suggest either inefficiencies in the rail transport sector or a lag in the economic payoff of broadband infrastructure investments. The negative impact of air transport (X_1) is relatively minor and statistically insignificant, possibly due to the country's regional focus in aviation and tourism dynamics. The constant term (+104,036) is relatively high, indicating the presence of other systemic economic contributors—such as remittances, natural resources, or state-led infrastructure projects—that are not directly accounted for in the model. In sum, the regression model offers strong insights into Egypt's digital and educational development as core contributors to economic progress. However, certain coefficients—particularly the negative values for transport infrastructure—warrant deeper analysis or model refinement to account for structural or sectoral nuances.

Indonesia

The obtained the regression equation for United Arab Emirates is:

$$Y = 0.00063X_1 + 0.00516X_2 + 4.76X_3 + 68.66X_4 + 0.95X_5 + 74.10X_6 - 84.41X_7 + 520932$$

The regression model for Indonesia shows remarkable explanatory power, with an R^2 of 0.999, indicating that nearly all variation in GDP per capita is captured by the independent variables. The F-statistic (1079.54) and its extremely low p-value (1.22E-07) confirm that the model is statistically significant. The normalized R^2 (0.83) and standard error (162.37) further attest to its robustness and accuracy.

From the equation, tertiary school enrollment (X_6) exerts the strongest positive influence on economic wellbeing, followed by internet usage (X_4) and railway passenger volume (X_2). These factors highlight the crucial role of education, digital connectivity, and mobility infrastructure in Indonesia's economic development. Conversely, secondary school enrollment (X_7) shows a negative coefficient, suggesting that while general education is essential, its current state may not be translating into direct economic gains—perhaps due to mismatches in labor market demands or educational quality issues. The broadband subscription variable (X_5) appears to have little effect, as its coefficient is near zero and statistically insignificant. Similarly, air transport (X_1) and mobile subscriptions (X_3), though with positive signs, have low statistical relevance in this model. The large constant term (+520,932) may reflect Indonesia's structural advantages or external economic factors not captured in the selected indicators—such as natural resources, trade patterns, or demographic dynamics. Overall, the model is highly accurate and provides useful insights into Indonesia's economic growth drivers—most notably, tertiary education and internet access—but would benefit from refinement to better interpret the role of basic education and telecommunication infrastructure.

South Africa

The regression line for South Africa is presented below:

$$Y = 0.00415X_1 + 0.05499X_2 + 8.97X_3 + 104.56X_4 - 156.24X_5 + 121.72X_6 - 137.43X_7 + 190441$$

The regression model for South Africa demonstrates a very high explanatory capacity, with $R^2 = 0.994$, meaning that nearly all variation in GDP per capita is explained by the model. The F-statistic (131.83) and significance level (0.0000229) confirm its statistical strength, while the normalized R^2 (0.82) and standard error (804.49) support the model's reliability. Among the variables, tertiary school enrollment (X_6) shows the most substantial positive contribution to GDP per capita, indicating the importance of higher education in South Africa's socio-economic progress. The coefficient for internet usage (X_4) also suggests a strong positive relationship with economic wellbeing, reinforcing the role of digital literacy and access. Interestingly, secondary education enrollment (X_7) and fixed broadband subscriptions (X_5) have negative coefficients. This may point to systemic inefficiencies or reflect challenges in the quality and applicability of general education and fixed internet services. The negative influence of broadband, in particular, could be due to limited penetration or affordability, which hampers its economic impact. Other variables—such as air transport (X_1), rail transport (X_2), and mobile subscriptions (X_3)—exhibit positive but statistically insignificant effects, possibly due to structural characteristics or low usage intensity in relation to economic output. The constant term (+190,441) implies the presence of other macroeconomic or structural drivers—such as natural resources, mining exports, or regional trade—that are not explicitly included in the model.

In summary, the model highlights the critical role of higher education and internet access in shaping economic growth in South Africa. However, it also reflects some inconsistencies—such as negative coefficients for education and broadband—which may require further investigation or model refinement to account for institutional, technological, or social context factors.

Iran, Islamic Republic

We developed a multiple regression equation for Iran, Islamic Republic:

$$Y = 0.0314X_1 - 0.0354X_2 + 44.77X_3 - 264.32X_4 + 899.51X_5 - 159.91X_6 - 18.52X_7 + 21{,}042$$

The regression model shows high explanatory power, with $R^2 = 0.949$, indicating that approximately 95% of the variance in GDP per capita is explained by the model. The F-statistic (15.93) and the associated significance level ($p = 0.00174$) confirm the overall statistical validity of the model. The adjusted $R^2 = 0.889$ and the standard error = 488.02 further support the model's reliability, particularly given the relatively small sample size (14 observations).

Among the independent variables: Air transport (X_1) has a positive and statistically significant effect ($p = 0.016$), emphasizing the importance of international mobility and logistics for Iran's economic performance. Secondary education enrollment (X_6) also shows a significant negative impact ($p = 0.017$), which may suggest structural inefficiencies or diminishing returns in the sector. Rail transport (X_2) and basic internet usage (X_4) exhibit negative coefficients, but only X_4 is statistically significant ($p = 0.025$), possibly reflecting underperformance or inefficiency in digital and physical infrastructure. Broadband subscriptions (X_5) show a positive effect, but the significance level ($p = 0.060$) is just above the conventional 0.05 threshold—indicating a potentially meaningful impact that warrants further investigation. Tertiary education enrollment (X_7) and mobile subscriptions (X_3) appear statistically insignificant in this model ($p > 0.09$), which may be due to either market saturation or complex indirect effects that are not captured linearly. The intercept (21.04) is moderate compared to other country models, suggesting that macroeconomic factors such as oil revenues, sanctions, and regional trade dynamics likely play a considerable role outside the scope of the modeled indicators. The regression results highlight the positive role of air connectivity and broadband access in Iran's economic performance, while also indicating inefficiencies in rail transport and parts of the education sector. These insights can be used to prioritize investment and policy reform, especially in the digital and infrastructure domains. Refining the model with additional variables—such as institutional quality, resource dependency, or regional trade—could enhance its interpretability and practical application for policymakers.

DISCUSSION

The obtained results demonstrate a significant impact of digital literacy and infrastructure on economic development and the reduction of transport inequality in BRICS+ countries. The conducted correlation analysis confirms the presence of a strong connection between the level of internet usage and GDP per capita ($R^2 > 0.85$), which indicates the crucial role of digital inclusion in economic well-being. Fixed broadband access indicators ($R^2 \approx 0.78$) turned out to be more significant predictors of GDP than mobile connections ($R^2 \approx 0.65$), emphasizing the importance of stable digital infrastructure.

Hence, the consolidated analysis of the influence exerted by three choosing blocks on prosperity and the reduction of digital inequality indicates that transport infrastructure—particularly railway transport—has a notably positive impact (exhibiting the strongest relationship with the Gini index). This finding suggests that, in large BRICS+ nations characterized by heterogeneous economies and urban population agglomerations (e.g., India, Brazil, China), the development of railway transport can provide a substantial impetus toward eliminating digital inequality across the entire country. Although education level and ICT service exports display weak negative correlations with the Gini index, this outcome indicates the potential role of education and digital technologies in mitigating inequality—albeit with currently limited effect, likely subject to further factors (investment, government programs, and regional disparities).

Furthermore, the study's findings convincingly demonstrate that the urban mobility block—encompassing Internet access and mobile or fixed broadband subscriptions—strongly influences GDP in the BRICS+ countries. Digital infrastructure (Internet coverage and mobile networks) shows the most robust positive correlations with per capita GDP in these nations, pointing to a direct link between economic growth and population engagement in contemporary digital services. The block responsible for digital literacy indicators (i.e., the share of the population with higher education) exhibits a significantly positive correlation with GDP, underscoring the importance of human capital in the digital development of BRICS countries.

Consequently, for BRICS nations, digital infrastructure and education emerge as the most crucial predictors of economic wellbeing (GDP), whereas inequality (Gini) is more closely tied to transport infrastructure. While digitalization and education possess high potential to reduce digital inequality and increase the economic wellbeing of populations, the current data indicate that their impact on the Gini index remains limited or ambiguous, necessitating more detailed investigation and consideration of each country's regional conditions.

Thus, the key factors in overcoming transport inequality and ensuring inclusive digital growth in developing countries are:

– high digital literacy of the population;
– stable broadband infrastructure;
– development of higher education (especially among women);
– balanced development of digital and transport logistics.

The obtained statistical relationships confirm that investments in digital development and human capital have a multiplicative effect on the wellbeing of the population and can become the foundation for sustainable and inclusive growth.

Based on our investigation, we found that in most BRICS+ countries, variables associated with internet penetration (individuals using the Internet) and mobile telephony (mobile cellular subscriptions) play a significant role. In some regression equations (e.g., for Russia and China), the fixed broadband subscriptions variable proves relevant. In the majority of models (particularly for China and India), access to the Internet and mobile connectivity emerges as a main driver of economic growth. In many cases (China, Russia, India), tertiary school enrollment exhibits a positive relationship with GDP, supporting the hypothesis that human capital (digital literacy, a qualified workforce) fosters economic development.

Regarding smart transport predictors, in most equations their effect is either modest or statistically insignificant. For instance, in China's equation, the coefficients for Air transport and Railways are positive, yet not the largest. In Russia and India, the sign often turns out to be negative or shows no effect. This leads to a strong, empirically grounded conclusion that transport infrastructure in certain BRICS+ countries can be a less clear-cut factor for improving economic wellbeing compared to the urban mobility and digital literacy blocks.

The study discovered certain similarities among BRICS+ countries. Urban mobility is crucial for economic growth in most countries where the model reached statistical significance (China, India, Russia). Higher education (X_6) generally exhibits a positive correlation with GDP in most countries, underscoring the importance of enhancing digital literacy as a driver for reducing digital inequality. In certain regressions (China, India, Russia), illogical or negative signs appear for some variables, implying the influence of additional factors or the short time span used, which hinders drawing robust causal inferences.

As for differences, in China one observes an extraordinarily strong impact of urban mobility's indicators (individuals using the Internet and so on), whereas in Russia, fixed broadband has greater influence and the model carries a large constant. In India, the most significant indicators also revolve around mobile services, but negative coefficients emerge for fixed broadband and the availability of Internet (possibly owing to unique growth dynamics or lagging effects).

The most reliable regression models (by formal criteria of R^2, the F-test, t-statistics, and MAPE) were for China and India. Russia (with $R^2 = 0.87$) also achieves decent results, albeit with a higher approximation error. Some countries (Brazil, Ethiopia) fail to produce statistically significant models (according to collecting data), suggesting that either the dataset is inadequate or the factors currently included do not explain GDP variability.

CONCLUSION

Empirical data on the development of digital, mobile, and transport infrastructure in BRICS+ countries aligns with our regression equations:

- China: Rapid expansion of mobile subscriptions and internet penetration (3G to 5G), large-scale investments in digital infrastructure and education. This matches the regression, where the digital block and higher education exert a strongly positive influence.
- India: Notable expansion of mobile telephony (e.g., affordable smartphones, "Digital India"), increasing internet penetration, but lags in infrastructure (especially fixed broadband). The model indeed shows a strong link with mobile indicators and a weaker relationship with fixed broadband.
- Russia: A relatively high proportion of tertiary education and broad accessibility of fixed broadband, as reflected in the model. Nonetheless, negative signs for some variables (internet users) might indicate complex geographic or economic factors.
- Iran: Known for difficulties in accessing global digital services (sanctions, constraints), which may account for the negative contribution of internet penetration (X_4). The higher education sector and air transport, areas of intensive Iranian investment (especially before tighter sanctions), will likely determine the country's digital development in the coming years.
- UAE: Mobile telephony and mobile internet may be displacing fixed broadband (X_5), hence why mobile connectivity and education are strong GDP drivers. The railway network is minimally developed, carrying little significance for enhancing the population's digital skills going forward.
- Brazil, Ethiopia and South Africa: No statistically reliable regressions are available, indicating that data collection needs further expansion and must account for the unique features of each country (primary exports, socio-economic stratification).

Thus, the actual dynamics of digitalization, educational coverage, and the expansion of mobile connectivity, combined with the positive coefficients in the regression equations, confirm that the urban mobility block and human capital plays a significant role in enhancing the overall resilience and technological advancement of smart cities in more than half of the BRICS+ nations. It means that research has allowed us to draw similar conclusions for emerging countries. Advanced intelligent technologies offer substantial advantages, yet their successful implementation requires a digitally literate population of emerging countries.

Accordingly, we have demonstrated that improving digital literacy in developing countries will contribute to fostering sustainable urban mobility for all citizens, eliminating inequalities based on socioeconomic status, race, or ethnicity. The development of intelligent urban transportation systems in major BRICS+ cities, alongside the use of digital technologies to enhance public literacy, will further facilitate the reduction of digital inequalities. At the same time, transportation-related variables (X_1, X_2) exhibit less stable effects and do not always attain statistical significance. Additionally, the presence of negative coefficients for secondary education (X_7) suggests the influence of unique national reform trajectories, underscoring the need for more detailed investigations

In conclusion the study findings confirm that digital literacy plays a pivotal role in reducing transportation inequality across BRICS+ nations. The study highlights that a digitally competent population has greater access to, comprehension of, and ability to utilize advanced mobility services. Particular emphasis is placed on the expansion of stable broadband infrastructure and mobile networks as key enablers of inclusive transport solutions.

The study also suggests that achieving near-universal Internet coverage can mitigate urban–rural disparities, especially in regions where conventional transportation systems, such as rail and air travel, remain underdeveloped or prohibitively expensive. Educational attainment, particularly tertiary education, emerges as another critical driver for BRICS+, as individuals with higher education levels tend to adopt new transportation technologies more readily.

Additionally, the findings indicate that robust logistical systems, including e-ticketing and real-time route planning, depend on widespread digital literacy for equitable adoption. Conversely, purely infrastructural approaches focusing on road, rail, and aviation networks may prove ineffective if substantial segments of the population lack internet access or the necessary digital skills.

The study underscores the significance of government-led initiatives, such as "Digital India" and similar programs in China and Russia (Digital Russia State program), in fostering both infrastructural and human capital development. In contexts such as the UAE and Iran, strong investments in mobile connectivity and higher education have already led to significant improvements in transportation accessibility. In contrast, the lack of comprehensive data for Brazil and Ethiopia limits the ability to draw definitive conclusions.

Overall, the research consistently emphasizes that human capital, as reflected in digital literacy and education, is fundamental in bridging regional transport gaps. As mobile payments and smart transportation applications continue to expand, the population's ability to engage with these digital tools will become increasingly crucial for ensuring equitable mobility outcomes. Looking ahead, public–private partnerships, expanded broadband coverage, and skill-building programs are identified as key mechanisms for reducing digital transport inequalities across BRICS+ nations. Finally, the findings highlight that measurable progress will depend on sustained policy support, continuous infrastructure investment, and ongoing monitoring of digital literacy indicators.

REFERENCES

Barns, S., Cosgrave, E., Acuto, M., & McNeill, D. (2016). Digital infrastructures and urban governance. *Urban Policy and Research, 35*(1), 20–31. DOI: 10.1080/08111146.2016.1235032

Butler, L., Yigitcanlar, T., & Paz, A. (2020). Smart urban mobility innovations: A comprehensive review and evaluation. *IEEE Access : Practical Innovations, Open Solutions, 8*, 196034–196049. DOI: 10.1109/ACCESS.2020.3034596

Camagni, R., Gibelli, M. C., & Rigamonti, P. (2002). Urban mobility and urban form: The social and environmental costs of different patterns of urban expansion. *Ecological Economics, 40*(2), 199–216. DOI: 10.1016/S0921-8009(01)00254-3

Ceder, A. (2020). Urban mobility and public transport: Future perspectives and review. *International Journal of Urban Sciences, 25*(4), 455–479. DOI: 10.1080/12265934.2020.1799846

Chen, Y., Wang, Z., & Ortiz, J. A. (2023). A sustainable digital ecosystem: Digital servitization transformation and digital infrastructure support. *Sustainability (Basel), 15*(2), 1530. DOI: 10.3390/su15021530

Costa, P. B., Morais Neto, G. C., & Bertolde, A. I. (2017). Urban mobility indexes: A brief review of the literature. *Transportation Research Procedia, 25*, 3645–3655. DOI: 10.1016/j.trpro.2017.05.330

Du, Z.-Y., & Wang, Q. (2024). Digital infrastructure and innovation: Digital divide or digital dividend? *Journal of Innovation & Knowledge, 9*(3), 100542. DOI: 10.1016/j.jik.2024.100542

Foltýnová, H. B., Vejchodská, E., Rybová, K., & Květoň, V. (2020). Sustainable urban mobility: One definition, different stakeholders' opinions. *Transportation Research Part D, Transport and Environment, 87*, 102465. DOI: 10.1016/j.trd.2020.102465

Gakenheimer, R. (1999). Urban mobility in the developing world. *Transportation Research Part A, Policy and Practice, 33*(7–8), 671–689. DOI: 10.1016/S0965-8564(99)00005-1

Grand View Research. (n.d.). *Smart cities market size, share & trends analysis report by application, by smart utilities, by smart transportation, by smart healthcare, 2024–2030: Market analysis report*. Grand View Research, Inc.

Greenstein, S. (2021). Digital infrastructure. In Glaeser, E. L., & Poterba, J. M. (Eds.), *Economic analysis and infrastructure investment* (pp. 409–452). University of Chicago Press., DOI: 10.7208/chicago/9780226800615.003.0009

Guo, B., Hu, P., & Lin, J. (2024). The effect of digital infrastructure development on enterprise green transformation. *International Review of Financial Analysis, 92*, 103085. DOI: 10.1016/j.irfa.2024.103085

Henfridsson, O., & Bygstad, B. (2013). The generative mechanisms of digital infrastructure evolution. *Management Information Systems Quarterly, 37*(3), 907–931. https://www.jstor.org/stable/43826006. DOI: 10.25300/MISQ/2013/37.3.11

Jan, B., Farman, H., Khan, M., Talha, M., & Din, I. U. (2019). Designing a smart transportation system: An internet of things and big data approach. *IEEE Wireless Communications, 26*(4), 73–79. DOI: 10.1109/MWC.2019.1800512

Jimenez, J. A. (2018). Smart transportation systems. In McClellan, S., Jimenez, J., & Koutitas, G. (Eds.), *Smart cities* (pp. 143–160). Springer., DOI: 10.1007/978-3-319-59381-4_8

Karami, Z., & Kashef, R. (2020). Smart transportation planning: Data, models, and algorithms. *Transportation Engineering, 2*, 100013. DOI: 10.1016/j.treng.2020.100013

Koutsikouri, D., Lindgren, R., Henfridsson, O., & Rudmark, D. (2018). Extending digital infrastructures: A typology of growth tactics. *Journal of the Association for Information Systems, 19*(10), 1001–1019. DOI: 10.17705/1jais.00517

Miskolczi, M., Földes, D., Munkácsy, A., & Jászberényi, M. (2021). Urban mobility scenarios until the 2030s. *Sustainable Cities and Society, 72*, 103029. DOI: 10.1016/j.scs.2021.103029

Mrityunjaya, D. H., & Kumar, N., Laxmikant, Ali, S., & Kelagadi, H. M. (2017). Smart transportation. In *2017 International Conference on I-SMAC (IoT in Social, Mobile, Analytics and Cloud) (I-SMAC)* (pp. 1–5). IEEE. DOI: 10.1109/I-SMAC.2017.8058235

Qin, X., Ke, J., Wang, X., Tang, Y., & Yang, H. (2022). Demand management for smart transportation: A review. *Multimodal Transportation, 1*(4), 100038. DOI: 10.1016/j.multra.2022.100038

Reddy, B. S., & Balachandra, P. (2012). Urban mobility: A comparative analysis of megacities of India. *Transport Policy, 21*, 152–164. DOI: 10.1016/j.tranpol.2012.02.002

Schipper, F., Emanuel, M., & Oldenziel, R. (2020). Sustainable urban mobility in the present, past, and future. *Technology and Culture*, *61*(1), 307–317. DOI: 10.1353/tech.2020.0004 PMID: 32249226

Serrano, W. (2018). Digital systems in smart city and infrastructure: Digital as a service. *Smart Cities*, *1*(1), 134–154. DOI: 10.3390/smartcities1010008

Stefansson, G., & Lumsden, K. (2009). Performance issues of smart transportation management systems. *International Journal of Productivity and Performance Management*, *58*(1), 55–70. DOI: 10.1108/17410400910921083

Tian, X., & Lu, H. (2023). Digital infrastructure and cross-regional collaborative innovation in enterprises. *Finance Research Letters*, *58*(C), 104635. DOI: 10.1016/j.frl.2023.104635

Van Wee, B., & Handy, S. (2015). Key research themes on urban space, scale, and sustainable urban mobility. *International Journal of Sustainable Transportation*, *10*(1), 18–24. DOI: 10.1080/15568318.2013.820998

Verma, B., Snodgrass, R., Henry, B., Smith, B., & Daim, T. (2020). Smart cities—An analysis of smart transportation management. In Tiwari, R., & Buse, S. (Eds.), *Managing innovation in a global and digital world* (pp. 401–422). Springer Gabler., DOI: 10.1007/978-3-658-27241-8_23

Wang, L., & Shao, J. (2024). The energy saving effects of digital infrastructure construction: Empirical evidence from Chinese industry. *Energy*, *294*, 130778. DOI: 10.1016/j.energy.2024.130778

Xu, H., Lin, J., & Yu, W. (2017). Smart transportation systems: Architecture, enabling technologies, and open issues. In Sun, Y., & Song, H. (Eds.), *Secure and trustworthy transportation cyber-physical systems* (pp. 13–26). Springer., DOI: 10.1007/978-981-10-3892-1_2

Yan, J., Liu, J., & Tseng, F. M. (2020). An evaluation system based on the self-organizing system framework of smart cities: A case study of smart transportation systems in China. *Technological Forecasting and Social Change*, *153*, 119371. DOI: 10.1016/j.techfore.2018.07.009

Zhao, C., Dong, K., & Taghizadeh-Hesary, F. (2023). Can smart transportation enhance green development efficiency? *Economic Change and Restructuring*, *56*(2), 825–857. DOI: 10.1007/s10644-022-09448-7

ENDNOTES

1 World Bank. World Development Indicators: Statistical Collection [Electronic resource] // World Development Indicators. — Access mode: https://databank.worldbank.org/source/world-development-indicators (accessed: 26.03.2025).

Chapter 5
Sustainable Practices in Nigeria's Mass Transportation:
Exploring Resources Allocation, Social Equity, and Government Policies

Sheneni Felix Maiyaki
https://orcid.org/0009-0006-1955-0637
Lagos Business School, Pan-Atlantic University, Nigeria

Frank Ojadi
Lagos Business School, Pan-Atlantic University, Nigeria.

Oluwasegun Raphael Odunlade
https://orcid.org/0000-0002-8351-8707
Swinburne University of Technology, Australia

ABSTRACT

This study explored the operations of mass transportation businesses in Nigeria that engage in inter-state trips and evaluated their sustainability based on resource efficiency and social equity. The study also delved into the role of government in mass transportation. This study adopted a qualitative approach. Seven mass transportation companies were selected using purposeful sampling, using the size of their operations as a selection criterion. Semi-structured interviews were conducted with terminal operators while the interview data was analyzed using thematic analysis.

DOI: 10.4018/979-8-3373-6335-6.ch005

Copyright © 2026, IGI Global Scientific Publishing. Copying or distributing in print or electronic forms without written permission of IGI Global Scientific Publishing is prohibited.

The findings indicate that mass transportation in Nigeria is bedevilled by a lack of awareness of sustainability, multiple taxation, heavy dependence on fossil fuels, and poor regulations as well as multiple taxations.

1. INTRODUCTION

Sustainability has become a global concern thanks to the UN Sustainable Development Goals. Several international agreements have been reached and numerous projects have been launched. Part of the SDGs is to make cities and human settlements inclusive, safe, resilient and sustainable. Under this goal is sustainable transport, which entails universal access, enhanced safety, reduced environmental and climate impact, improved resilience, and greater efficiency of transportation systems (UN, 2021). The need for the actualization of this goal cannot be overstated because transportation aids the flow of resources from regions of abundance to areas of scarcity. Hence, it is imperative to work towards the shift from traditional transportation systems to sustainable ones.

Achieving sustainability requires a holistic approach (Sachs, 2015; Elkington, 1997). Existing literature indicates the existence of three pillars of sustainability, which are environmental, economic and social considerations. In furtherance to this, Gunasekaran et al. (2014) opined that it is pertinent that any sustainable practice adopted by businesses should meet the requirements of environmental regulations and improve business performance. Satisfying the three sustainability conditions entails carefully deploying sustainable practices (Wang et al., 2012). Therefore, sustainable transportation should be perceived as safeguarding the environment, promoting stakeholders' welfare and guaranteeing the business's financial performance. The latter emphasizes business continuity and its relationship with stakeholders of the transportation sector (Litman & Bruwel, 2006), which prompts the establishment of indicators of sustainable transport (Zhou, 2012; Singh et al., 2020).

In Nigeria, inter-state transportation is chiefly done using mass transportation countries. Road and air transport are the available means of transportation connecting the regions of Nigeria. However, given the high poverty rate in Nigeria, air transportation is utilized only by a few. Mass transit companies' adherence to the three sustainability pillars needs to be be facilitated. Poor transportation has been a bane to the development of the nation. The waste of agricultural produce and developmental imbalance are among the many repercussions of the poor flow of people and materials. There is a need to take a deep dive into this transportation sector to bring to light their operational practices and find areas of improvement.

Unfortunately, sustainability is also not a well-discussed topic outside of academia in Nigeria. Hence, the absence of a road map, measurement metric, operational model, or systematic approach needed to attain sustainability in Nigerian sectors (Oraegbune, 2022). The availability of these is pertinent to guaranteeing that businesses welcome sustainability advocacy and begin to adopt sustainable practices. In the global sphere, researchers and corporations are now interested in how various business operations and processes can be more sustainable to enhance organizational competitiveness and profitability (Wang et al., 2012). It is imperative to replicate this in the sectors of individual countries, especially in developing countries where sustainability does not feature in business operations. This begins with research that evaluates the sustainability of various business processes and operations. Researchers and corporate bodies should synthesize and assess the conformity of their operations to multiple dimensions of sustainability (Oraegbune, 2015).

Furthermore, poverty, institutional deficiencies, poor education, and technological setbacks are prevalent in developing countries like Nigeria. Given these, there needs to be more research on the unique conditions of businesses operating in Nigeria and how the shift to sustainability can be achieved. This is more so in the road transportation sector, which contributed 45.1% of the CO_2 emissions attributed to transportation in 2018 (Our World in Data, 2020).

Some scholars have conducted studies on sustainable transportation. Oraegbune (2022) conducted a study on sustainable road infrastructure delivery in Nigeria where the researcher investigated the theory and practice of sustainable practice adoption in road and infrastructure delivery in Nigeria. On the other hand, Chikwendu et al. (2015) researched sustainable road transportation in Nigeria, focusing on problems associated with carbon emissions. Abam et al. (2021) also conducted a study on the environmental sustainability of the Nigerian transportation sector with an emphasis on ways of saving energy in the Nigerian transport sector and reducing carbon emissions to ensure ecological sustainability. Most of the focus has been on environmental sustainability, whose good ideals and prospects will likely be adopted by businesses if the economic and social sustainability prospects are catered to.

Consequently, this research explored the operational practices of mass transit businesses in Lagos-Nigeria that engage in inter-state transportation from the perspective of resource efficiency and social equity, evaluate them, and prescribed ways to facilitate a sustainability shift. Knowing how integral government policies and regulations are, this study also inquired about government policies and how they impact the operations of mass transportation companies.

Hence, the questions this research sought to answer are:

1. What are the resource allocation practices of mass transportation companies in Nigeria?
2. How is social equity reflected in the operational practices of mass transportation companies in Nigeria?

3. How do government policies and regulations impact the sustainability of mass transportation?

2. LITERATURE REVIEW

Mass Transportation in Developing Countries

Sound transportation systems are integral to fostering the liveability of communities (Miller et al., 2013). It ensures the ease of movement of people and commodities. Ease of transportation is necessary for the overall development of countries. Interestingly, most developing countries' populations rely on public transportation (Guillen et al.,2013). However, the number of private cars is on the rise year on year. Nigerian road transportation comprises private cars, minibusses, and traditional buses used for mass transit and private purposes (Chukwendu et al., 2015). Traffic congestion, which is chief among the problems faced in Nigeria and other developing countries, is caused by the high amount of private cars, which can be attributed to the loss of confidence in the public transport service. This has led to inconsistent arrival and departure times, the primary cause of customer dissatisfaction (Mutiawati et al., 2022).

There have been many studies on sustainable transportation in developing countries. Mansoor et al. (2022) studied the factors affecting non-motorized transport and the benefits of non-motorized transportation and addressed future challenges for developing countries. The research found that improved health and environmental benefits were key benefits of non-motorized transport. The research recommended a great need for stakeholders in developing nations to grasp the factors affecting the use of non-motorized transport considering their socio-economic condition and investigate the benefits of non-motorized transportation. The study also recommended non-motorized transportation as an integral part of the solution to road transportation challenges.

Similar studies have been conducted to assess adopting the globally recommended sustainability drivers in developing countries. Jianhong et al. (2021) researched the impacts of car-sharing on vehicle ownership in Shangai. Hinging on the global recommendation that car-sharing reduces private car ownership, the study used operation data and questionnaires to determine if car-sharing impacts vehicle ownership. The study found that car-sharing has suppressed the growth of private cars by reducing individual willingness to purchase a car. Car sharing has also succeeded in parts of Europe and North America. The researchers recommended that efficient transport systems and car-sharing services are necessary for reducing private car ownership. Similarly, Saxena and Gupta (2023) conducted a study that compared the demographic and travel characteristics of existing carpoolers and those of two-wheelers and single-occupant car users in Gurugram, India. The study stated that car ownership is a sign of economic prosperity. The research found similar characteristics in car-poolers and single-occupant car users, suggesting that single–occupant car users are potential carpoolers.

Another drawback is the need for standard metrics for evaluating sustainability in mass transportation. Naim et al. (2022) affirmed the difficulty of assessing the performance of public transportation services in developing countries due to the unavailability of reliable data and the need for formal guidelines. The researchers recommended a proposed inventory and framework involving data envelopment analysis, enabling policymakers and city planners to provide improved transportation services to the general public. This study reveals a critical problem of sustainable transportation. There are hardly institutional policies and measurement metrics to facilitate the sustainability of transport systems in Nigeria.

To address the problem of measurement metrics, Daimi and Rebai (2022) conducted a research titled "Sustainability Governance Indicator-based Framework for Public Transport Companies in Developing Countries." The study aimed to create a sustainability governance framework that assesses public transport companies in developing countries. The study suggested four critical performance indicators viz environmental, economic, social, and institutional indicators. After analyzing three transportation companies in a developed country, the study recommended an implementation framework comprising the four sustainability indicators using an additive weighing method.

Understanding Sustainable Transportation

Sustainability is the continuous fulfillment of the economic needs of the current generation in ways that are not to the detriment of future generations (Themes et al., 2004). Sustainability can also be seen as economic growth that is environmentally sustainable and socially all-encompassing (Sachs, 2015). At its origin, sustainability hinged on intergenerational equity. As seen in the Brundtland Report (1987), sustainable development was viewed as satisfying the present needs without stopping future generations from meeting theirs. However, it has now ultimately encompassed environmental integrity, social welfare and economic needs regardless of time and location (Litman & Burwell, 2006). This concept has evolved to one that is more centralized, encompassing economic, environmental and social considerations for everyone (Sachs, 2015), now and in the future.

Understanding sustainable transportation requires an understanding of transportation. Transportation has one primary objective: moving people, equipment and supplies safely and efficiently from one location to another (Gurtu et al., 2019).

Litman (2007) opined that transportation constitutes the movement of people and goods and is one of the integral movers of economic development and quality of life. Given the far-reaching potential of adequate transportation, its sustainability must be guaranteed. Little wonder that the nature, scope and implications of sustainable transportation have gained traction (Litman & Burwell, 2006). Global concerns about climate change, energy utilization and resource management have led to the sustainability evaluation of different sectors of the economy; the transport system, which is not left out, requires new and unique approaches (Hill, 2009).

Nevertheless, there is a continuous increase in negative economic, environmental and social impacts with the constant growth of motorized transportation (OECD, 2002).

Unsustainable practices are very evident in road transportation. The transportation sector accounts for 25% of global CO_2 emissions (The Maritime Executive, 2020). CO2 emission has increased exponentially due to the number of vehicles that ply roads (Singh et al., 2020). This is further aggravated by most vehicle's dependence on fossil fuels. Gurtu et al. (2016) stressed that transportation companies and vehicle manufacturers should consider alternative power sources. The transportation sector also causes noise pollution. Moving vehicles generate both engine sounds and rolling noise, contributing to noise pollution and causing vibration in the surrounding et al., 2020). Many ailments, such as high blood pressure, can be caused or triggered by noise beyond 70 decibels. Similarly, road accidents have devastating effects on individuals, families and society, which affect individual income, productivity at work and losses to insurance companies (Elvik, 2020), not to mention overwhelming the health sector. One consequence of unsustainable transportation is traffic congestion. Traffic causes employees to resume late to their offices and consequently reduces work delivery efficiency, hampering the sustainability of other systems (Singh et al., 2020).

An integral part of human survival in the 21st century is sustainable mobility. Hence, countries must adopt sustainable practices in which low-cost transport is integral. It is also important that regions of countries are more connected to foster overall national development. To achieve this, sustainability must be integrated into the decision-making process. It means that environmental, economic, and social considerations are factored into the decision-making process in the transportation sector. Researchers have stressed that sustainable decision-making is planning that considers goals and the impact of business processes no matter how hard they are to measure (Litman & Burwell, 2006). One major challenge in sustainable transportation is balancing environmental, economic and social dimensions in decision-making (She et al., 2012). Achieving the triple bottom line of sustainability can be quite challenging when the transportation business caters to the needs of different income groups (May et al., 2001).

Contrary to popular opinion, sustainability is concerned with systems analysis, not threat analysis (Transportation Research Board, 1997), to ensure the capacity for long-term continuity (Center for Sustainability, 2004). This entails designing business operations with sustainability in mind. If this is achieved, many benefits abound. Sustainable transportation is less expensive and should meet the needs of all sections of society (Singh et al., 2020). It is a chief factor in contributing to sustainable development. Evidently, to guarantee sustainable development, the rate of consumption of both renewable and non-renewable resources should not exceed the regeneration rate (Daly, 1991). Gurtu (2014) affirmed that the pollution rate should be the same or less than the time the environment requires to purify itself.

The sustainability of mass transportation is closely linked with customer loyalty. Vincente et al. (2020) found a positive relationship between public transport companies' commitment to sustainability and passenger loyalty. This finding corroborates that of the European Commission (2008) and Kennel (2016), who found that customers refer to environmentally sustainable solutions even if they are more expensive than unsustainable alternatives. Haghshenas and Vaziri (2012) conducted a study that assessed the sustainability performance of transportation systems in 90 cities across various countries using a composite index comprising of nine indicators, three each for the sustainability pillars. The economic indicators were transportation expenditure, average user cost and average time spent in traffic. The social indicators were transportation-related fatalities, accessibility and variety of transportation means. The environmental indicators were emissions, energy consumption and land consumption. Their findings indicated that the developed parts of Europe and Asia performed better than other cities due to more emphasis on mass transportation, cycling and walking.

In evaluating the sustainability of mass transportation companies, many indicators abound. Sadly, traditional transportation systems focus on vehicle mobility and traffic congestion performance metrics, neglecting sustainability indicators (Zeng et al., 2013). Doing transportation the traditional way may increase the negative consequences of climate change (Zanni & Bistrow, 2010), contrary to the need for sustainable transport to meet the needs of all sectors of society, which is necessary for sustainable development (Singh et al., 2020). Transporters should, however, adopt safety, comfort, and social interaction processes to ensure social sustainability. Economic sustainability can be achieved if transportation costs are minimized through speed optimization, increased vehicle capacity and reduced commuting time (Singh et a, 2020). The transport research community agrees that mass transportation systems are more sustainable than private transportation systems due to their relative social, economic and environmental performance (Litman, 2009; Currie & De Gruyter, 2018).

Sustainability is evaluated using a set of indicators to monitor trends, assess policies and prescribe performance targets (Litman, 2003). Sustainability indicators showing the best pointers to steer human activities toward sustainability differ from country to country due to varying challenges and priorities (Kumar et al., 2015) calling for standardization. Kennedy (2005) opined that effective governance on land use; efficient, fair and adequate funding; strategic infrastructure investment and proper neighborhood design are sustainability indicators. Sustainable transportation implies efficient energy use (Sigh et al., 2020). Most studies suggest social, environmental, and economic indicators for assessing sustainability (Liyanage et al., 2017). The indicators in Table 1 offer a more comprehensive set of indicators for assessing the operational sustainability of mass transportation using the tipple bottom line. This study has adopted a few of them due to the absence of quantitative data in Nigeria's transportation sector as affirmed by Naim et al (2022).

Table 1. Indicators for assessing the operational sustainability of mass transportation businesses

INDICATOR	UNITS	DESIRABILITY
Environmental		
Energy Consumption	MJ/pkm	Less is desirable
Fuel Consumption	L/pkm	Less is desirable
Pollutants Emission	kg	Less is desirable
Noise Pollution	dB	Less is desirable
Land area consumed by transport facilities	m2	Less is desirable
Water usage	L	Less is desirable
Ecological Impacts of Right of Way	m	Less is desirable
Social		
Accessibility	%	More is desirable
Cumulative opportunity	jobs/activity centres	More is desirable
Average user trip distance	km	Less is desirable
Affordability	fare/income per capital	Less is desirable
Populations exposed to transportation-related emissions	Number of people	Less is desirable
Transportation-related fatalities	Average annual fatalities	Less is desirable
Transportation-related accidents	Average annual accidents	Less is desirable
Economic		
Annual operating cost	N/km	Less is desirable
Overall system capital cost	N	More is desirable

continued on following page

Table 1. Continued

INDICATOR	UNITS	DESIRABILITY
Individual route capital cost	N	Less is desirable
Cost recovery	%	More is desirable
Cost Subsidies	%	Less is desirable
Average financial cost per trip	N	Less is desirable
Average time cost per trip	hours	Less is desirable

Source: Adapted from De Gruyter et al. (2017) and Currie & De Gruyter (2018) and modified by the authors.

There have been a couple of research targeted at the sustainability performance of mass transportation systems. Miller (2014) developed a framework comprising twenty indicators grouped into economic, environmental, social and system effectiveness categories for assessing the sustainability performance of thirty-three rail systems in selected cities in Asia and the Middle East. Miller et al. (2016) further developed the study of Miller (2014) by evaluating the range of data normalization and indicator-weighing techniques for appraising the sustainability performance of different mass transportation modes for the Broad Corridor in Vancouver.

De Gruyter et al. (2017) adapted Miller's framework to appraise the sustainability performance of urban mass transportation systems in selected cities in Asia and the Middle East.

3. THEORETICAL BACKGROUND

Triple Bottom Line

The triple bottom line which provides a sustainability measurement framework for businesses was coined by John Elkington in 1997. This construct has become synonymous with sustainability (Alhaddi, 2015). Elkington postulated that for businesses to be sustainable, three perspectives are necessary, namely, environmental, economic and social perspectives. The sustainability of a business can be ascertained through the assessment of these bottom lines. Henriques (2007) enthusiastically affirmed that TBL is a brilliant and far-reaching metaphor. The reason why is clear. TBL expands the environmental protection agenda by integrating the economic and social dimensions (Elkington, 1997).

Figure 1. The triple bottom line

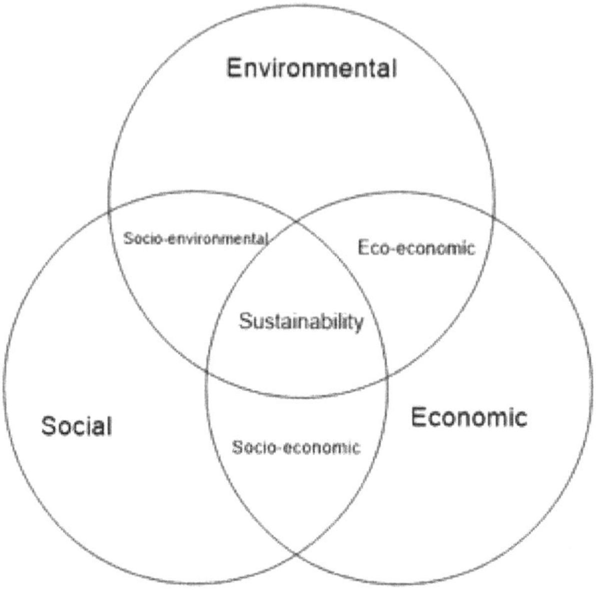

Source: *Adapted by the authors from Miller (2020)*

Environmental Perspective

The environmental sustainability perspective seeks to safeguard natural capital (Elkington, 1997). It stresses that business activities should not compromise the environment's quality and future generations' ability to cater to their needs. Businesses should use resources efficiently, minimize the emission of greenhouse gases and reduce their ecological footprint(Goel, 2010).

Social Perspective

The social perspective of TBL is considered the people aspect. It refers to ensuring that the practices of businesses are fair and beneficial to the stakeholders of the business (Alhaddi, 2015). This involves the acquisition of virtues such as loyalty, honesty and dependability, which brings about enormous benefits for the organization because of low levers of social friction (Elkington, 1997). It emphasizes the interaction of the organization and the community and how the organization engages its stakeholders (Goel, 2010).

Economic Perspective

The economic perspective of TBL borders on how the operations and practices of the organization impact the organization financially and the economy in general. It emphasizes the result of the organization's activities on the economic system (Elkington, 1997). As one of the aspects of sustainability, this perspective concerns how businesses and the economy at large can survive and advance to support future generations (Spangenberg, 2005; Alhaddi, 2015).

Two perspectives arose regarding the three pillars of sustainability: weak and strong. Weak sustainability merges the three pillars and can tolerate the neglect of one pillar so long as the value of the other pillars being upheld creates more value than the harm the negligence of the others causes (Constanza & Daly, 1992). For instance, focusing on the socioeconomic aspect of sustainability and neglecting the environmental implications is considered a worthwhile tradeoff, provided the level of environmental degradation does not outweigh the benefits generated by the socioeconomic aspects. This perspective, which is based on making tradeoffs, faced criticism. The environment should not be sacrificed for economic gains and environmental degradation cannot be measured in monetary terms (Nasrollahi et al., 2018), nor can the social dimensions be measured in monetary terms.

Strong sustainability holds the environment in the first place. Environmental sustainability is considered to have the capacity to pave the way for social and economic dimensions of sustainable development (Taghvaee et al., 2023). Research suggests that contrary to the concept of weak sustainability, trade-offs in issues of sustainable transport is unsustainable (Litman & Burwell, 2006). At the center of the discussion of sustainable transportation is achieving the three core pillars of sustainability without trade-offs under the minimum benchmarks (Litman & Burwell, 2006). The ideal attainment of sustainability is found in integrating all three pillars of sustainability (Nikolaou & Tsalis, 2018; Tagvaee et al., 2023). Due to these merits, this study has adopted TBL as its theoretical foundation.

4. RESEARCH METHODOLOGY

This section outlines the research approach, data collection methods, and analytical techniques employed to investigate the operational sustainability of mass transport businesses in Nigeria. This study draws upon the Triple Bottom Line (TBL) framework coined by Elkington (1997), which encompasses three key dimensions: economic, environmental, and social.

Research Approach

The research adopts a qualitative approach to gain in-depth insights into the sustainability practices of mass transportation companies in Nigeria. Qualitative research is deemed appropriate for this study as it allows for a nuanced exploration of the intricate factors influencing sustainability within the context of the Nigerian transportation industry (Creswell & Poth, 2018).

Sample Selection

The sample for this study comprises terminal managers of seven top registered mass transportation companies with terminals located in Lagos, Nigeria. The selection of Lagos as the primary location is strategic, considering its status as a major economic and transportation hub within the country. Terminal managers were chosen as respondents due to their critical roles in overseeing and shaping the operational practices of these companies.

Additionally, the inclusion criteria stipulate that selected companies must transport people to more than one region of Nigeria and make an average of three trips daily. This criterion ensures that the study encompasses a specific range of national mass transport businesses. By focusing on registered companies, we aim to establish a foundation of credibility and accountability within the sampled population.

Data Collection

To collect primary data, semi-structured interviews were conducted with terminal managers of the selected mass transportation companies. The semi-structured nature of the interviews provided flexibility in exploring various issues related to resource allocation and social equity. It allowed for probing questions and enabled participants to share their perspectives, experiences, and insights (Denzin & Lincoln, 2018).

Data Analysis

Thematic analysis will serve as the primary method for data analysis in this study. This approach systematically identifies, analyzes, and reports patterns (themes) within the interview data. Themes will be derived through a rigorous process of data familiarization, generating initial codes, searching for themes, reviewing themes, defining and naming themes, and finally, writing up the analysis (Braun & Clarke, 2006).

Indicators of Operational Sustainability

Resource Allocation

Energy Consumption: Assessment of energy consumption patterns within mass transport companies.
Fuel Consumption: Examination of fuel usage and its impact on sustainability.
Water Consumption: Analysis of water usage and its implications on resource sustainability.

Social Equity

Coverage: Evaluation of the geographic reach and accessibility of mass transport services.
Frequency: Examination of service frequency to determine its impact on user convenience.
Reliability: Assessment of the dependability and consistency of transport services.
Affordability: Analysis of fare structures and their affordability for different socioeconomic groups.
Inclusivity: Investigation into the inclusivity and accessibility of transport services for diverse populations.

By employing the thematic analysis method and focusing on these sustainability indicators, we aim to gain a comprehensive understanding of the operational sustainability practices of mass transport businesses in Nigeria. This research design allows for an in-depth exploration of the complexities and challenges these companies face in aligning with the principles of the Triple Bottom Line framework.

5. FINDINGS AND DISCUSSION

Resource Allocation

In exploring the resource allocation in the operations of Mass transportation companies in Nigeria, several questions were asked to this effect. Terminal managers were asked questions about power generation and usage in the terminals, fuel consumption by their fleet of buses, and water sourcing and usage. It was found that the primary source of power at the terminal is premium motor spirits (PMS). Inverters powered by generators (or solar panels in some cases) are used by a few terminals. This is evident from the following comments made by some of the respondents.

"There is electricity which you know is nothing to write home about in Nigeria. That is why we have to use generators because to pump water, you need electricity. We have to power the generator very early in the morning. As early as 3:30 am/4 am"

Similarly, another respondent mentioned that:

"For electricity, we use a generator, we also have an inverter, also we make use of the NEPA (power company)"

Another commented that:

"Given the increase in fuel price, we use solar panels to supplement the cost of running a generator. However, when there is no power and the solar is down, we still use the generator. We also use the generator to charge the batteries."

In justification of using generators, a respondent stated:

"Well, for this terminal, we do not have solar. For the power company, it is anytime you see it. It is not peculiar to this place. In my house, for five months I have been running the entire building on solar and generator because the transformer has been down. So in the case of this place, you can never predict power supply"

Then in explaining power availability at the terminals, a terminal manager emphasized:

"We always have electricity here. When the power company does not supply power, we use our generator. This is because we want to maintain a constant power supply in the terminal. Generator or power company"

Secondly, the researchers found that all the buses use PMS. This further indicates Nigeria's high dependence on fossil fuels for mass transportation. Sadly, Mass transportation companies have not considered any possibility of reducing dependence. They seem to think they are left to their fate. The Mass transportation companies also do not track the usage of fuel. They use the drivers' opinions to estimate the fuel budget. The faulty nature of this resource allocation method is seen in the disparities between the fuel budget for the same trip by various Mass transportation companies. The study also found that Mass transportation companies practice letting the buses go on a trip with fewer passengers than the bus can carry so long as they do not incur a loss (in a bit not to delay the passengers). These can be seen in the following comments made by the respondents:

"I do not think the fuel price can be reduced. If the drivers are moving from one destination to another, they need petrol. In fact, when the bus is not full, we still give the drivers the same money that we give them when the bus is full. They need fuel so there is not much we can do... It is business, we gain today and may lose tomorrow. And during the festive seasons, the price goes up. So we still make our money"

Similarly, a terminal manager said

"I do not think we can reduce the consumption of petrol. It is based on the journey and there is no way to reduce it"

Another respondent said:

"I do not really know. When the driver is leaving we give them a fixed amount of money for fuel but for the liters, I don't know the consumption level"

In line with the practice of letting the buses ply the road with fewer passengers than it can contain, some comments were:

"Previously, when we have between five to seven passengers, we let them go. But now, due to the high cost of fuel, we try to get nine and above passengers. This is because if we do contrary, we will not make any profit when we close the books. We need profits to continue the business. But we cannot determine the response of passengers...Once we see that the money we get from ticket sales is enough to settle the driver and have a little money in the purse, we dispatch the bus immediately"

Similarly, another respondent commented:

"We have a fixed departure time. We do our best to stick to it. However, sometimes passengers may delay and we need to wait for them. But if the delay is too much, we just go like that. We delay because if the number of passengers in a bus is small, we may not make a profit"

The implication of constant usage of PMS at the terminals on the sustainability of Mass transportation companies is far-reaching. Constant use of PMS increases the operating cost of the terminals not forgetting the role of continuous use of fossil fuels on the ecological footprint. It is however relieving to see that some of the businesses are already making sustainable decisions in using solar panels to generate electricity for use. Although using generators occupies the prime position, this is a move in the right direction. It can also be seen that the unreliability of the electricity companies is a key factor in promoting the use of fossil fuels. This finding aligns with the study of Chirisa et. al (2015) on Africa's transportation and the environment. The study found that poor fuel price structures, over-reliance on private cars and heavy dependence on fossil fuels as characteristics of Nigeria's transportation system. Mass transportation systems are preferred to private modes of transportation because they have less emission per passenger (Dirgahayani, 2013), occupies less space (Chirisa et. al, 2015), and causes less traffic (Litman, 2016). If buses are then allowed to ply the road with less than their capacity, this will diminish the benefits of mass transportation.

Regarding water usage, it was found that many of the terminals take responsibility for generating water for themselves. They have boreholes that cater to their water needs. This is because whenever the electricity company does not provide electricity, the generators run the pumping machines. The study also found out that the terminal managers do not track water usage. Water is used and replenished as needed. Water conservation is achieved through regulations that encompass rationing water and recycling. It requires the use of specific technologies (Kurunthachalam, 2014). Unfortunately, water conservation practices is not common among Mass transportation companies in Nigeria. The following comments were made in this light.

"We pump water. We pump water constantly. When there is power supply, we pump water and when the power goes off we use generator... Once we check the tank and there is no water, we begin to pump"

Another mentioned that:

"For the water, we have tap water, we have toilet facility... We have a borehole that supplies the tap with water"

Another interesting comment was:

"Water is just directly generated by us because we have a well"

Government Policies & Regulations

The study also sought to find the impact of government policies and regulations and lack thereof on the operations of Mass transportation companies. Various issues were brought to bear through the responses of the terminal managers. First, mass transportation operators perceive a lack of support and interest from the government. The study found that numerous government agencies interact with Mass transportation companies mainly for taxation purposes.

A manager lamented that

"The individuals who have invested in it are not being encouraged to the best of my knowledge. For us, we have been in the business for 30 years. But wherever we are now is as a result of self-struggle the government is not really helping there is really no government support."

Another comment with regards taxation and charges was

"The charges I can say that they are necessary. But they are excessive. When you are on the road you pay, when you come to the terminal, you pay. Like here we pay parking permit. We pay the Local government and the State government will come separately. Then on the highway, the federal government. We pay all these at every given time. We have to pay LASMA, VIO, Road Safety, Police and what have you. So when you look at it the charges is enormous"

Public-private partnerships and collaborations are integral to the success of every sector of a nation (Moeliodihardjo et. al, 2012). The absence of this collaboration would stifle growth. It is therefore pertinent that the government of Nigeria to seek ways of collaborating with the transportation sector and unifying their taxation systems.

This study also found that the lack of regulations on those who can engage in mass transportation has led to competition from unregistered transporters. However, the registered transporters are preferred. The fuel subsidy removal has also adversely affected the transport companies. These can be seen in the following quotations from the interview transcript.

With regards the regulations and unregistered transporters, the following was said:

"Really the regulatory side of it is nothing to write home about that is why it is like an old comers business. You see someone who is retired from his office may be he has a minivan he was using as his private car when are looking for ways to make ends meet and Put it by the side of registered mass transportation company's terminal. Then when the registered transportation company are charging for instance or 10,000 for instance to give a standard service. They can collect 7,000 naira because they are trying to make ends meet it doesn't matter to them. The owner of the minivan or maybe the driver or the person he entrusted the car to collects lower amounts and anything goes. Because there is no proper regulatory system."

However, a terminal managers had this to say:

"The thing is that the unregistered transporters want to outshine but most customers prefer use registered transport companies because of the insecurity in the country. They are traceable when they use registered transport companies and something happens along the way. We have had cases where people sue us when unsatisfied with our service. This cannot happen with unregistered transporters. This is because we are a legal entity. And the customers have more confidence in the registered transporters"

Similarly, another terminal manager mentioned that:

"Most people prefer to use the registered transporters. It is when were are fully booked then they go elsewhere. They don't mind the price"

These comments were made with regard to the effects of the removal of fuel subsidy removal

"Prior to the removal of fuel subsidy, we usually load seven to eight buses. On the days of eastern sit-at-home, we load four to five buses. But since the removal of the fuel subsidy, there has been less patronage. The highest we do is six and during the sit at home we load three or four"

Another respondent said:

"Less people are travelling now and also we have competitors. The unregistered transporters charge less than we do. Those who necessary have to travel explore cheaper alternatives"

Another mentioned that

"So the price of petrol went up, we also increased our prices and patronage went down."

Competition such as the one faced by Mass transportation companies in Nigeria has a negative impact on business performance. Business performance which is the level of a company's achievement of its targets (Purnama & Subroto, 2016) is impacted when competition comes from businesses that have less operating cost. The tendency is for registered Mass transportation companies to reduce service quality to compete with the unregistered transporters of to target a different economic class which will mean poor access to standard transportation to people with low income. Another threat is the transporters' safety when they are transported by unregistered transporters across regions of the country.

Social Equity

In exploring the social equity consideration of the Mass transportation companies critical questions bordering on coverage, frequency, reliability, affordability, inclusivity and access among others were asked. Observations around the social sustainability practices of Mass transportation companies reveal a high level of indigenous methods towards achieving social equity and inclusivity. In this regard, the study's findings showed that stakeholder considerations are at the forefront of business and corporate decisions despite profit intentions.

For instance, the study revealed that most of the Mass transportation companies pre-inform their customers of the departure time through calls. There are further indications that announcements are made through the MTB's Public Address System to communicate the departure time of a bus. As related by a manager of one of the Mass transportation companies sampled.

Thus evident from this finding is the level of customer-centric nature of these businesses. However, it can be deduced that, Vulnerable groups (VGs) such as passengers with hearing difficulties might be excluded. This observation could have severe operational implications. For instance, a terminal manager noted that

> *"In the morning, we resume by 5am, we go to our system, we check on those people that booked online a day before. We either will call them to confirm if we are picking them up or they are coming. Then people start coming in and we begin to sell tickets to them. After selling tickets to them, they have different route to go, first bus leaves at 5:45am. We announce the bus leaving, then they go out to board the bus."*

However, it was equally observed, that while inclusivity may not be practiced in the dissemination of information, it seems to be a prominent theme in the travel management of customers with different needs. Observations revealed a strong indication of the practice of social inclusivity that recognizes differences in socio-economic classes and the unique needs of VGs among the different Mass transportation companies that were sampled. This is because most Mass transportation companies provide different types of buses to cater for different socio-economic groups. Although the price difference is not very significant and it comes with a lower service level. Furthermore, inclusivity for the vulnerable groups was significantly revered among all Mass transportation companies evident in the provision of designated seats and preferences. The first response in this regard shows the minor deficiencies in handling VGs.

".....when we have them like maybe one person. And when you say people with disabilities, you do not mean someone with no legs, a blind person, for instance, will have to travel with someone to care for him. To help them eat and ease themselves. But in the absence of that, we will talk to the driver if the person is going on a straight journey. But not that we have a section for people with disabilities"

However, another respondent said

"....Some of them come with their relatives and some the driver takes care of them till they get to the destination terminal. There is also seat number six. It has more leg space and it is close to the door. It is reserved for them".

Another response that shows considerable provision for VGs was that

": I have had them many times. Even yesterday, I had to deal with people with disability. We give them special care. We do not let them feel like it is a big deal. We do not know the future. Only God. Some of them were normal before. Some became disabled due to accidents. So we handle them with courtesy and respect. Depending on the level of the disability. Someone who is blind needs a relative to travel with them. If they do not, they sit at seat number two next to the driver. The driver takes care of them. For those who are deaf or cannot speak, I make them my friends. We communicate by writing for those who can read and write."

Findings of the study also revealed a moderate understanding of the needs of vulnerable groups especially the physically impaired by the Mass transportation companies. Although there are no indications that they have formal training in this regard, which could further explain the deficiencies in spreading this across all business operations, nonetheless they adopt unique methods to ensure inclusiveness and equity for the VGs. To this end, suggestions and recommendations were made in subsequent sections of this study to improve their practice of social inclusivity. Moreover, the provision of different pricing mechanisms to cater to different affordability categories. A respondent related how there are luxurious provisions for economic groups that can afford it and a reduced-cost alternative for lower socio-economic groups.

"Our prices depends on where the person is going to. Our CEO have given all the customers a 30% discount so if we do not have enough money we cannot fuel the bus. We charge less than other companies. We have AC bus and non-AC bus. The non-AC bus is one thousand naira less."

Evident also from the findings of the study is the thinking around the social well-being and quality of life of the external and internal customers of the businesses. On the part of external stakeholders (passengers), the study revealed that Mass transportation companies make arrangements to pick up passengers at locations convenient to them rather than having them move to designated parks. This would have implications on the additional resources in terms of time, environment, and stress that could be incurred. Demonstrated instances of good corporate governance within the top echelon of the business hierarchy leading a move that prioritized the social aspect of sustainability by ensuring optimal service provision in low passenger turnout scenarios, pickups, and stopovers for food during travel. These social equity efforts are indications of prioritizing sustainability beyond profitability.

As related by a Manager,

"We pick up passengers on the way. The only condition is that they should only have a backpack. This is because we do not want delay. We have a fixed departure time. We do our best to stick to it. However, sometimes passengers may delay and we need to wait for them. But if the delay is too much, we just go like that. We delay because if the number of passengers in a bus is small, we may not make a profit."

The implication of this as revealed in the findings of this study is that the external customers have a sense of belonging and loyalty to the business. This is evident as these passengers or stakeholder customers, in reciprocating, also recognize this, influencing them to make pro-business decisions. These observations strengthen the position of previous studies that when the public as a stakeholder in a business is well treated in terms of the ESG components, they see the business as an extension of themselves and are influenced to make decisions to ensure its sustainability. This helps sustain customer loyalty. Moreover, observations revealed that Business Branding, identity, culture and legal registration give a sense of security and promote brand confidence and awareness. Furthermore, the provision of customer bio-data by most Mass transportation companies indicates a concern for the livelihood and security of the passengers.

Furthermore, findings along this line revealed that most of the Mass transportation companies communicate travel delay challenges to the passengers. The influence of this as observed in this study is the recognition of good customer relations during delays and this is an indication of robust stakeholder management. While some of these methods are non-conventional, they are indigenous and widely practiced, which suggests that African businesses have indigenous and culturally defined sustainable practices, which may not wholly align with international practice but have a good reflection of global best practices. These indigenous sustainable practices and their impact on business operation in the unique African context makes for future research exploration.

It was observed that there were considerable efforts to cater for other groups of employees with a flexible work schedule. The Mass transportation companies make available bespoke flexible working schedules that prioritize employees' well-being and work-life balance. This speaks to the people or stakeholder orientation of the Mass transportation companies. While the work schedule is an early resumption and late closure type, the flexible work arrangement makes work-life balance easy for the employees.

"After the buses leave in the morning, the staff can go wherever they want. Only a few remain to attend to customers who want to book in advance"

Passengers and employees are not the only stakeholders for which a sustainable business operation is responsible. The business equally has responsibility to the host community. In this regard, the study showed that a minority of the Mass transportation companies try to have a responsibility for the host community. According to a respondent, they provide water for their use.

"We pump water. We pump water constantly. When there is power supply, we pump water and when the power goes off, we use the generator."

A respondent has this to say about how they supply the communities with water.

"No, we do not. We use it for the toilets and when there is power supply, others who trade around come to fetch water here too."

While this is a bold move to ensure water is continually available for use, it brings to the fore the issue of ineffective resource allocation, given that the pumping of water is continually done, without due consideration for the volume used per day per terminal. These businesses do not monitor water usage although they make water available for the public, and stakeholders, an indication of the prioritization of the S in ESG even maybe without knowing and ensuring the livelihood security of their host community. This could have implications for energy resource efficiency. Moreover, a lack of synergy was observed between the Mass transportation companies reflecting a lack of maximization of the External Economies of Scale. When asked if it is possible to collaborate with other similar Mass transportation companies in instances of low patronage, the response was,

"No. We do not collaborate with other transporters. Only among our terminals."

These observations could have added implications for energy cost per unit because of operational efficiencies and synergy. This implies that these businesses require knowledge by way of recommendations on alternative and sustainable practices that could suffice as a contribution to the host community.

6. CONCLUSIONS AND RECOMMENDATIONS

This study has explored the practices of mass transportation businesses in Nigeria in order to have an understanding of its sustainability. It highlights the operations of mass transportation businesses, bringing to light the sustainability (or unsustainability) of their operations as well as the role of government policies. As an exploratory study, it intends to kickstart the conversation on how the Nigerian transportation sector can achieve the sustainability shift from an environmental, economic, and social perspective.

From the findings of this study, the following are recommended.

1. Mass Transportation Businesses should
    i. Adopt more sustainable sources of energy like solar energy. They may also use natural liquified gas and electricity to power their buses.
    ii. Track their resource usage in order to eliminate wastage where possible.
2. Government and Its Agencies:
    i. Set regulations to guide the industry to guarantee the adoption of best practices. These regulations should encompass environmental guidelines, safety standards, fees, and levies.
    ii. The government should carve out clear jurisdiction for taxes and levies collection from mass transportation companies. A centralized payment system should also be adopted.
    iii. The government should promote sustainable transportation through its agencies and incentivize the shift from traditional operations to sustainable ones.
3. News Agencies:
    i. It is pertinent that the media presents sustainability in the light of the triple bottom line to help businesses and policymakers see the benefits of sustainable alternatives.

REFERENCES

Abam, F. I., Ekwe, B. E., Diemuodeke, O. E., Ofem, M. I., Okon, B. B., Kadurumba, C. H., Archibong-Eso, A., Effiom, S. O., Egbe, J. G., & Ukueje, W. E. (2021). Environmental sustainability of the Nigerian transport sector through decomposition and decoupling analysis with future framework for sustainable transport pathways. *Energy Reports*, *7*, 3238–3248. DOI: 10.1016/j.egyr.2021.05.044

Alhaddi, H. (2015). Triple bottom line and sustainability: A literature review. *Business and Management Studies*, *1*(2), 6–10. DOI: 10.11114/bms.v1i2.752

Braun, V., & Clarke, V. (2006). Using thematic analysis in psychology. *Qualitative Research in Psychology*, *3*(2), 77–101. DOI: 10.1191/1478088706qp063oa

Chirisa, I., Bandauko, E., & Gaza, M. (2015). Transport and the Environment: A Critical Review of Africa. *Zhongguo Renkou Ziyuan Yu Huanjing*, *13*(4), 309–319. Advance online publication. DOI: 10.1080/10042857.2015.1078489

Creswell, J. W., & Poth, C. N. (2018). *Qualitative Inquiry and Research Design: Choosing Among Five Approaches*. Sage Publications.

Daly, H. E. (1991). *Steady State Economics*. Island Press.

Denzin, N. K., & Lincoln, Y. S. (2018). *The Sage Handbook of Qualitative Research*. Sage Publications.

Dirgahayani, P. (2013). Environmental Co-benefits of Public Transportation Improvement Initiative: The Case of Trans-Jogja Bus System in Yogyakarta, Indonesia. *Journal of Cleaner Production*, *58*, 74–81. DOI: 10.1016/j.jclepro.2013.07.013

Elkington, J. (1997). *Cannibals with forks – Triple bottom line of 21st-century business*. New Society Publishers.

Elkington, J. (1997). *Cannibals with Forks: The Triple Bottom Line of 21st Century Business*. Capstone Publishing.

Goel, P. (2010). Triple bottom line reporting: An analytical approach for corporate sustainability. *Journal of Finance, Accounting, and Management*, *1*(1), 27–42.

Gurtu, A. (2014). A Framework for Reducing Greenhouse Gas (GHG) Emissions through Carbon Pricing for Offshore Outsourcing. (Doctoral dissertation, Ryerson University, Toronto). Retrieved from http://digital.library.ryerson.ca/islandora/object/RULA%3A3388

Henriques, A. (2007). CSR, sustainability and the triple bottom line. In Henriques, A., & Richardson, J. (Eds.), *The Triple Bottom Line: Does it All Add Up?* (pp. 26–33). Earthscan.

Hill, C. (2009). Transportation and sustainability best practices background. *Proceedings, AASHTO Sustainability Peer Exchange*, Washington, DC, 32.

Kumar, R., Mahdu, E., Dahiya, A., & Sinha, S. (2015). Analytical Hierarchy Process for Assessing Sustainability: Indicators of Public Transportation Systems, Pedestrians and Feeder Services in Developing Countries. *World Journal of Science, Technology and Sustainable Development,* 12*(4).

Litman, T. (2007). Developing indicators for comprehensive and sustainable transport planning. *Transportation Research Record: Journal of the Transportation Research Board, 2017*(1), 10–15. DOI: 10.3141/2017-02

Litman, T. (2009). Sustainable transportation indicators: A recommended research program for developing sustainable transportation indicators and data. Paper presented at the 88th Transportation Research Board (TRB) Annual Meeting, Washington, DC.

Litman, T. (2016). *Evaluating Public Transportation Health Benefits.* Victoria Transport Policy Institute.

Litman, T., & Burwell, D. (2006). Issues in sustainable transportation. *International Journal of Global Environmental Issues*, *6*(4), 331–347. DOI: 10.1504/IJGENVI.2006.010889

Liyanage, C., Dias, N., Amaratunga, D., & Haigh, R. P. (2017). Current Context of Transport Sector in South Asia: Recommendations towards a Sustainable System. *Built Environment Project and Asset Management*, *7*(5), 490–505. DOI: 10.1108/BEPAM-10-2016-0051

May, A. D., Page, M., & Hull, A. (2001). Developing a set of decision-support tools for sustainable urban transport in the UK. *Transport Policy*, *15*(6), 328–340. DOI: 10.1016/j.tranpol.2008.12.010

Miller, K. (2020). The Tripple Bottom Line: What it is & Why it is Important. *HBS Online.* Accessed via: https://online.hbs.edu/blog/post/what-is-the-triple-bottom-line

Miller, P. (2014). *Sustainability and public transportation: Theory and analysis* (Doctoral dissertation). Calgary, Manitoba: University of Calgary.

Miller, P., de Barros, A. G., Kattan, L., & Wirasinghe, S. C. (2016). Analyzing the sustainability performance of public transit. *Transportation Research Part D, Transport and Environment*, *44*, 177–198. DOI: 10.1016/j.trd.2016.02.012

Mutiawati, C., Suryani, F. M., Anggraini, R., & Isya, M. (2022). Importance-Performance Analysis in Public Transport Level of Service: A Case Study of the Tran Koetaradja Bus in Banda Aceh. *Aceh International Journal of Science and Technology (Banda Aceh)*, *11*(1), 70–84. DOI: 10.13170/aijst.11.1.23146

Naim, N. H., Adnan, M. S. G., Dewan, A., & Zannat, K. E. (2022). Assessing the Performance of Public Transport Services in a Developing Country: A Case Study Using Data Envelopment Analysis. *Growth and Change*, *53*(1), 377–409. DOI: 10.1111/grow.12588

Nasrollahi, Z., Hashemi, M., Bameri, S., & Taghvaee, V. M. (2018). Environmental pollution, economic growth, population, industrialization, and technology in weak and strong sustainability: Using STIRPAT model. *Environment, Development and Sustainability*, *22*(2), 1105–1122. DOI: 10.1007/s10668-018-0237-5

Nikolaou, I. E., & Tsalis, T. (2018). A framework to evaluate eco- and social-labels for designing a sustainability consumption label to measure strong sustainability impact of firms/products. *Journal of Cleaner Production*, *182*, 105–113. DOI: 10.1016/j.jclepro.2018.02.042

OECD. (2002). *OECD Guidelines Towards Environmentally Sustainable Transport*. Organization for Economic Co-operation and Development.

Oraegbune, M. O. (2015). Appraisal of Transportation Infrastructure Sustainability in Nigeria.(Ph.D. Thesis, Department of Civil Engineering, Faculty of Engineering, University of Nigeria, Nsukka).

Oraegbune, O. M. (2022). Towards Sustainable Road Infrastructure Delivery in Nigeria. *Arid Zone Journal of Engineering. Technology & Environment*, *18*(2), 319–330.

Purnama, C., & Subroto, W. T. (2016). Competition Intensity, Uncertainty Environmental on the use of Information Technology and its Impact on Business Performance of Small and Medium Enterprises. *International Review of Management and Marketing*, *6*(4), 984–992.

Report, B. (1987). Report of the World Commission on Environment and Development: Our Common Future. *United Nations*. Retrieved from https://sustainabledevelopment.un.org/content/documents/5987our-common-future

Sachs, J. D. (2015). *The Age of Sustainable Development*. Columbia University Press. DOI: 10.7312/sach17314

Saxena, A., & Gupta, V. (2023). Carpooling: Who is Closest to Adopting it? An Investigation into the Potential Car-poolers among Private Vehicle Users: A Case of a Developing Country, India. *Transport Policy*, *135*, 11–20. DOI: 10.1016/j.tranpol.2023.03.007

She, S., Lu, Q., & Ma, C. (2012). A probability–time&space trade-off model in environmental risk perception. *Journal of Risk Research*, *15*(2), 223–234. DOI: 10.1080/13669877.2011.634515

Singh, A., Gurtu, A., & Singh, R. K. (2020). Selection of Sustainable Transport Systems: A case study. *Management of Environmental Quality*, *32*(1), 100–113. DOI: 10.1108/MEQ-03-2020-0059

Spangenberg, J. (2005). Economic sustainability of the economy: Constructs and indicators. *International Journal of Sustainable Development*, *8*(1/2), 47–64. DOI: 10.1504/IJSD.2005.007374

Taghvaee, M. V., Arani, A. A., Nodehi, M., Shirazi, J. K., Agheli, L., Ghojogh, H. M. N., Salehnia, N., Mirzaee, A., Taheri, S., Saber, R. M., Faramarzi, H., Alvandi, R., & Rahbarian, H. A. (2023).. . *Review of Economics and Political Science*, *8*(2), 134–161. DOI: 10.1108/REPS-12-2019-0168

The Maritime Executive. (2020). Transport uses 25 percent of world energy. Retrieved from https://www.maritime-executive.com/article/transport-uses-25-percent-of-world-energy

United Nations. (2021). Sustainable transport, sustainable development. *Interagency report for the Second Global Sustainable Transport Conference.*

Vicente, P., Sampaio, A., & Reis, E. (2020). Factors influencing passenger loyalty towards public transport services: Does public transport providers' commitment to environmental sustainability matter? *Case Studies on Transport Policy*, *8*(2), 627–638. DOI: 10.1016/j.cstp.2020.02.004

Wang, X., Chan, H. K., Yee, R. W., & Diaz-Rainey, I. (2012). A two-stage fuzzy-AHP model for risk assessment of implementing green initiatives in the fashion supply chain. *International Journal of Production Economics*, *135*(2), 595–606. DOI: 10.1016/j.ijpe.2011.03.021

Zhou, J. (2012). Sustainable transportation in the US: A review of proposals, policies, and programs since 2000. *Frontiers of Architectural Research*, *1*(2), 150–165. DOI: 10.1016/j.foar.2012.02.012

Chapter 6
Blending Routes and Fares (Fees) as a Tool to Enhance Mobility Using Various Types of Urban Transport in St. Petersburg

Egor Rianovich Temirgaliev
https://orcid.org/0000-0002-5344-2000
Peter the Great St. Petersburg Polytechnic University, Russia

Zoya Simakova
https://orcid.org/0000-0002-8462-7358
Peter the Great St. Petersburg Polytechnic University, Russia

Nikita Lukashevich
https://orcid.org/0000-0001-5719-4844
Peter the Great St. Petersburg Polytechnic University, Russia

Fazliddin Khujaev
Tashkent State University of Economics, Uzbekistan

ABSTRACT

The urban transport infrastructure is an intricate multistage system. In many cases, this system is in a state of continuous advancement and also enlarge convenience. In most cases, each time consumers interact with system, they mix miscellaneous

DOI: 10.4018/979-8-3373-6335-6.ch006

Copyright © 2026, IGI Global Scientific Publishing. Copying or distributing in print or electronic forms without written permission of IGI Global Scientific Publishing is prohibited.

ways of transport when building routes. Due to the rapid elaboration of modern technologies, a large quantity of data has already been accumulated, processed, and made available in an easy-to-use form. Treatise investigates the global undergo of creating such transport systems. Similar characteristics and prerequisites for usage of various approaches in the implementation of projects have been pointed up. A resemblance of the urban transport infrastructure of some megacities and St. Petersburg has been conducted. Amount of residents in St. Petersburg is constantly increasing, and it poses new challenges to the city's transport infrastructure. In this regard, the scientific paper offers some information tools that allow analyzing existing routes and elaborating new ones.

INTRODUCTION

Public transport in a contemporary city is very essential for its success, because it enables to create the city a more convenient place to live, reduces the time spent on trips and cargo conveyance. The model of the United States, where it is impossible to move in cities without a car has shown that this method of people's movement is irrational. American cities, despite the many wide highways in the middle of cities occupy 5 out of 10 places among the cities with the largest traffic jams in the world. At the same time, European cities with narrow streets and practically no wide avenues compete with them in this regard. This is due to the so–called "fundamental law of traffic congestion", which was discovered by the American economist Anthony Downs back in 1962. It sounds like this - the wider the road, the more cars there are on it. In this regard, many European cities have begun to base their planning not on the capacity of streets in cars, but on its ability to accommodate the quantity of people, and public transport is the undisputed leader. It is not enough just to construct a public transport system, it is important to create one that would be convenient for city residents and that would not require them to travel by car for daily trips. In this scientific paper, it is proposed to analyze the public transport system of European cities, consider changes in the Moscow system, as well as a feasible change in the public transport system in St. Petersburg (Duranton & Turner, 2011).

METHODOLOGY

Analysis of Existing Transport Systems

Analysis of Berlin's Public Transport System

Let's take a closer glance at Berlin's public transport system. Appendices A and B provide complete diagrams of the public transport system. It is on this basis that the redesign of the public transport system in St. Petersburg will be based in this treatise. This choice was made because of the similar history of the cities. Both cities were mainly built up from the 17th to the 18th century, both were capitals of their respective states and for this reason both had similar buildings and road systems in their current historical centers. Then, in the late 19th and early 20th centuries, cities rapidly industrialized and overgrown with suburbs and industrial areas. This was followed by the era of totalitarian regimes and war, which greatly damaged the cities. The subsequent urban reconstruction after World War II and the beginning of panel's housing mass construction. The only difference at this stage is that at that time the city was divided into 2 parts, and West Berlin developed a little differently from its eastern part, so mainly consider East Berlin to determine the most accurate patterns in the construction of the transport system. In order to show by instance what the traffic pattern in Berlin looks like, the authors chose the Pankow district, which includes next districts:

- 0301 Prenzlauer Berg
- 0302 Weißensee
- 0303 Blankenburg
- 0304 Heinersdorf
- 0305 Karow
- 0306 Stadtrandsiedlung Malchow)
- 0307 Pankow
- 0308 Blankenfelde
- 0309 Buch
- 0310 Französisch Buchholz
- 0311 Niederschönhausen
- 0312 Rosenthal
- 0313 Wilhelmsruh (JLL, 2017)

The transportation system of this district is stemmed from the U2 line and the C8 line. The ring lines of the S41 and S42 machines also pass through it (they duplicate each other in opposite directions). Authors decided to get a view through the Alexanderplatz area, through which there are many clubs (U2, U5, U8) that are prohibited (S3, S5, S7, S9), and also the Reichsban (RE1, RE2, RE7), which go to neighboring cities such as Brandenburg, Cottbus, Bad Belzig, therefore it will play a significant role in the district system overall (Berliner Verkehrsbetriebe, n.d.),

From this station, the district includes two tram lines M2 and M4. Buses in the Berlin's transport system act as district transport or transport that brings S–ban, U-ban and trams to the main line, where electric trains and subways are superfluous. Trams also most often either duplicate metro and commuter train stations between branches or connecting them with each other using transfers. The routes do not duplicate each other for a long time, most often intersecting at several stops for transfers and most often near metro and commuter train stations. Most often, routes are quite long, performing several functions at once – communication between districts, transportation to mainline transport.

The transport system is derived from the principle of routes' minimal duplication. There is duplication, but mostly on small sections of routes, and this mainly applies to trams for a logical reason – this transport is limited to its own tracks. At the same time, it is smooth to get from one area to another due to transfers, which take only 5-10 minutes, since all routes have an interval of 5 to 15 minutes, with the rare exception of 20 minutes on suburban routes to areas of private development. And at the same time, passengers do not have to pay for each boarding, since the city focuses on one-time travel tickets, but on long-term travel, which makes public transport more attractive from an economic point of view for passengers, and small intervals grant to get to any point in the city without adjusting to the transport schedule, since the passenger has confidence that he will leave the bus stop within 5-10 minutes (Live Streets, 2013).

The city's ticket system itself is as follows: the city is divided into 3 zones – A, B and C.

Tickets can be purchased for each zone individually, or for several at once, which makes the ticket much more expensive. Tickets by zone apply to all types – for one-time trips and for a certain period of time. A ticket for a one-time trip does not work the way the authors are used to seeing in St. Petersburg. It does not work for one trip in one type of transport, but for an unlimited amount of transfers within 2 hours, which enables to actively use the built-in transport system, which involves fast transfers.

Apart from all, most of the routes run along streets where there are indicated lanes for public transport. This increases its predictability, as intervals are not disrupted and transport capacity increases, as it does not stand in traffic jams. However, it should be borne in mind that indicated lanes are not made everywhere, but only in those places where public transport has traffic problems.

Helsinki Public Transport System Analysis

Helsinki repeats many aspects of the Berlin transport scheme, but with its own peculiarities. Since, unlike Berlin, Helsinki is a city on many islands, this imposes certain restrictions in the form of "bottlenecks". And this issue was circumvented by creating switching routes. Images 6,7,8,9 show the traffic pattern of Helsinki trams, as well as how routes are switched. This technique enables to keep the traffic interval low in dense buildings, and at the same time not create a bunch of short routes, and passengers can get with fewer transfers or without them at all to the desired address, even on the other side of the city. Naturally, the point about the absence of transfers can be relevant only if the city is not too large, otherwise the intervals in a number of traffic options will be too long. But in the case of St. Petersburg, this can be practiced with satellite cities, which are connected to the city by one or two highways, the same bottlenecks.

Prague Public Transport System Analysis

The Prague system is similar in its construction principles to Berlin's, and it is presented in Appendix B. There is also no duplication of routes, the emphasis is on rail transport. Nevertheless, there are many more captivating solutions for marking streets in Prague.

Since the streets in this city are very narrow, urban planners have to dodge a lot to be able to arrange everything necessary to move around the city. Such experience could be useful for creating a transport system in a number of St. Petersburg central districts, outside the main avenues. To get on or off a tram, people essentially have to jump under cars, and in heavy snowfalls, instead of trampled snow, there is sleet on the road. This is extremely unsafe and leads to pedestrian collisions, as can be seen on the accident map. From 2015 to 2024, the only reported cases in this area for this reason were 7 accidents (out of 10 accidents) with minor or severe injuries. This is on a section less than 100 meters long. For comparison, we can give an example of the intersection of Ligovsky Ave and Candle Alleyway. At the same time, 18 accidents occurred there, of which only 6 were pedestrian collisions. And this is provided that in the first example there is no pedestrian crossing, but there is one on Ligovsky, which further indicates the need for platforms for safety and comfort (Dopravni podnik hlavniho mesta Prahy, n.d.)

Vienna Public Transport System Analysis

Using the example of Vienna, the researchers would like to analyze the convenience of transfers from one transport to another. This aspect is vitally important in the public transport system. As mentioned earlier, in a big city, no mode of transport can fully operate alone.

Therefore, it is important not only to plot routes so that they intersect, but also to make transfers that would take a minimum of time for passengers. I would also like to mention an important but inconspicuous aspect. The street in this place was made curved, which leads to a natural decrease in traffic speed, which means increased road safety in an area of increased pedestrian congestion. Partially, this project was done correctly by placing one bus stop in a triangle, next to the entrances to the subway and the construction of a new pedestrian crossing. But on the other hand, the tram stop remains 137 meters from the nearest metro entrance. The same thing happens with the second bus stop. It was left in its original place, instead of being moved to this triangle.

Thus reducing the journey and time for passengers by 85 meters. The sample of a nice hub in St. Petersburg is Ladozhsky Railway Station. It enables to get from any type of transport to another in less than a couple of minutes, and commuter trains, metro, tram and bus converge there. One of the most atrocious from the passenger's point of view is Kupchino. In order to get from the subway to commuter trains, you need to go down to the underpass and climb back up to the commuter train platform, and this is the best part of this hub. Since in order to get from the metro station to the bus stop from the Frunzensky district, it is necessary to overcome 250-300 meters, which is 5 minutes at a steady pace, which is too much for an interchange hub. But there is an option to fix such a hub, although this entails high financial costs. First of all, it is necessary to place the suburban train tracks not away from the metro station, but on the sides of the platform, as is done at Devyatkino station, which are identical in design to Kupchino station. In this case, the railway tracks are shifted to the west, and there is a place for the organization of the tram and bus rings closer to the entrances to the underpass, which also mix to the west as unnecessary. Thus, it turns out to reduce the transfer time from the metro to commuter trains to 20 seconds, and the path to ground transport will be only 100-120 meters, instead of 250-300. This significantly increases the attractiveness of the transfer, and consequently the attractiveness of transport in general, while practically not affecting the capital commercial development. This may not seem significant, but it all relates to one thing – the comfort of public transport for passengers. At the moment, Vienna's public transport is considered one of the most comfortable in Europe. Back in the 60s, the residents of Vienna themselves joked about their public transport, calling it transport 3A - Alcoholiker, Arbeitslose und Asoziale (alcoholics, unemployed and antisocial). Quote from the book - Verkehr ist kein Schicksal. In den 1960er Jahren entstand der Witz: Wer fährt noch mit dem öffentlichen Verkehr? Die Drei A: Alkoholiker, Arbeitslose und Asoziale [Traffic is not fate. In the 1960s, the joke arose: Who still uses public transport? The Three A's: Alcoholics, Unemployed and Anti-Social] (Knoflacher, 2017)

The previously mentioned point about reducing speed near a crowd of pedestrians is also important, since, according to observations, a comfortable environment for pedestrians greatly increases the effective range of a public transport stop. People find it convenient to get to a bus stop if it runs along a street that is car-oriented in 100% of cases only 50 meters away, and below 50% this value drops already at 150 meters.

At the same time, in a car-free zone (meaning a zone where there are no cars or they can move at a low speed), in 100% of cases, people consider a 200-meter path to be comfortable, and below 50% this value drops to about 375 meters, which is equal to the average distance between stops. This aspect is also major because it is essential to understand what kind of urban environment prevails on the streets and how much bus stops are capable of covering the area around them. And an uncomfortable urban environment, aimed primarily at car traffic inside the city, itself provokes an even more uncomfortable environment due to the formation of traffic jams. Therefore, in the same Vienna, they actively worked with this so that public transport would work, and not drive by itself. And, for example, they made some division of streets into a conditional classification – transit roads, of which there are few in the city, and they are primarily needed to collect traffic and its subsequent exit to highways, both within urban and suburban areas. And in this context, it is important to understand that a motorway is a road that does not run on the same level as the building and the rest of the streets. There are dozens more aspects, from comparing the speed of movement along a route at different types of stops to using free–flowing streets, but it will not be possible to display all of them in this work. Between 1993 and 2012, the percentage of car usage decreased from 40% to 27%. Then authors predicted a decrease to 20% (Prillinger, n.d.).

Over the entire period of modernization of Vienna's transport system, the following results have been achieved The relationship between the specific use of public transport and the level of accidents in cities. Vienna has significantly improved its position since 1991.

In this city, as in many others, the “Zero Vision” program is used. Vision Zero is an international program to improve road safety and reduce deaths in road accidents. One of the main principles of Vision Zero is that road users will sometimes make mistakes, and infrastructure should never make mistakes.

Every time there is a fatal accident on a city street involving a car, an investigation is conducted and the factors that led to the accident are clarified. Changes are being made to the infrastructure. This is achieved by different methods, as each case is analyzed separately.

Moscow Public Transport System Analysis

In recent years, public transport reforms have been actively conducted in Moscow, and among the most famous is the creation of the MCD and BKL – urban electric trains, which were not only launched with a short interval (6-10 minutes), but also actively integrated into the city's transport system, thereby unloading metro lines and eliminating the need for many suburban buses. In addition, the launch of the Moscow Central Diameters in 2019 unloaded 19 stations of the Moscow metro by 12%. There is much less duplication of routes, the intervals have been reduced to 10-15 minutes on most routes, and alphabetic indexing of routes has appeared.

This indexing allows local residents to quickly navigate. As an experiment, an "antikarman" was made on Pyatnitskaya Street. This method of placing a bus stop is suitable for use on streets where parallel parking is allowed for two reasons:

- it is easier and faster for buses to stop by eliminating unnecessary maneuvers. Despite the existing regulations on allowing people to leave a public transport stop, many drivers do not do this, both according to various observations and personal experience;
  - this type of bus stop allows you to organize more parking spaces than a classic bus stop.

Unfortunately, this type of stop is not yet fixed in GOST R 52766-2007 and for this reason it is used only as an experiment. Although, according to the response from Rosstandart, this GOST does not contradict the possibility of creating this type of stopping points.

Common Features, Differences, and Their Causes

All these European cities have the following similarities:

1. Avoiding duplication of routes All transport is based on the principles according to which the metro was built in Russia – 1 way – 1 route. Of course, it is not always possible to comply with this, and sometimes it is necessary to retreat on sections of several kilometers if there is a very large passenger traffic, and on other streets it is either impossible to transport or if it is impractical, but the key here is that this is the exception, not the rule.
2. An expensive one–time ticket is a cheap subscription Here is a comparison of ticket prices for different cities (AB zones)
   - Berlin one-time ticket 2.40 euros -29 euros per month
   - Helsinki 3.2 euros - 60.10 euros per month

- Prague 32 CZK - 550 CZK per month
- Vienna 2.40 euros – 51 euros per month
- Moscow 70 rubles – 3450 rubles Petersburg 70 rubles - 3665 rubles (does not include clocked commuter trains).

The difference in cost by city is as follows

- Berlin 12 times
- Helsinki 17 times
- Prague 21 times
- Vienna 21 times
- Moscow 49 times
- St. Petersburg 52 times.

This aspect helps to reduce the overall amount of cars on the streets very well, as it makes regular transportation more attractive in cost.

3. Approximately equal intervals on different routes. In many places in Moscow, this rule now also works. The interval ranges from 10 to 15 minutes, which enables to make transfers without much loss in time, which means it allows you to better connect different types of transport and improve connections in the city. In St. Petersburg, on many routes, the interval ranges from 15 to 30 minutes. To understand the effectiveness of using land transport in two Russian cities, one can compare their population and the number of rolling stock for land transport - in St. Petersburg the population is 5,597,763 people, in Moscow 13,149,000 people. There are 7,500 rolling stock in St. Petersburg and 9,100 in Moscow (Rambler, 2024).

The difference in population is 2.35 times, and in the transportation – 1.21, which is a perfect indication of the effectiveness of even the imperfect Moscow system. Transfers enable to increase the frequency of movement. Let's imagine a simplified model of a city consisting of three main residential areas and three main points of activity. If we set the task of organizing a public transport system for this city, then the first impulse will be to take a direct route from each district to each of the activity centers. We will have a system consisting of 9 routes.

CREATING A MODEL FOR CREATING A PUBLIC TRANSPORT SYSTEM

List of rules for creating a public transport system in a modern city.
The rules in brief:

1. Orientation of the system's construction away from mainline transport;
2. Convenient transfer hubs located next to each other, and preferably in one place, but at different levels, especially if the node is large. The best example of such a hub in St. Petersburg is Ladozhsky Railway Station.;
3. Small intervals of ground transport movement from 5 to 15 minutes within the city, depending on passenger traffic, instead of the average values for the city from 15 to 30 minutes;
4. Make branching routes from the suburbs if passenger traffic is low;
5. Don't take short routes;
6. Routes should not be isolated, there should be as many possible transfers as possible, both between routes of the same type of transport and completely different types.;
7. The possibility of purchasing travel tickets in different locations;
8. The intervals during the day should not be very different, otherwise it leads to a decrease in the attractiveness of public transport;
9. Convenience of transport boarding points, convenience and comfort of approach to them;
10. A travel card for a certain period of time must be valid for all types of transport, including commuter trains within the clock range, as already introduced in Moscow at the IDC;
11. All modes of transport should work as a whole, not separately.;
12. Prioritization of transport, which determines the distance between stops
    - Railway – 2-3 km;
    - Metro 1-1.5 km;
    - Tram 500- 700 meters;
    - Bus and trolleybus 200-300 meters.
13. Mainline transport in the form of railways and subways should not duplicate the existing road network, while trams can sometimes connect different areas in new locations using flyovers. That is, the task of the mainline transport is to increase the connectivity of the city or districts.;
14. Creating a division into streets primarily for public transport and pedestrians and focused on personal transport, but not creating full-fledged highways from them;
15. Create dedicated lanes for public transport where necessary.

The main framework of the system is the metro and city trains Given the possibility of transforming the city's railway system.

Zone A can be made within the half-ring of the railway (Lakhta – Ladozhsky – Ugolnaya Gavan).

Zone B will be everything within the Ring Road + adjacent settlements – Bugry, Murino, Novoe Devyatkino, Yanino, Novosaratovka, Shushary, Novogorelovo, Novoselye.

Zone C – approximately within the boundaries of the projected St. Petersburg Ring Road 2.

Zone D – distant suburbs – Tosno, Vyritsa, Sosnovy Bor, Shlisselburg, Zelenogorsk, Roshchino. Made by analogy with Helsinki and Berlin There are also many railways inside the city that are used only for industrial zones, but at the same time they could be used as an alternative to the subway, having previously upgraded them or raised them above street level, or made them shallow or simply open below street level. Such possible sections for railway communication within the city – Utkina Zavod – Kudrovo- Murino-Toksovo Vsevolozhsk – Kushelevka – Lakhta Vsevolozhsk – Ladoga – Coal Harbor/Krasnoe Selo/ Gatchina/Pavlovsk Toksovo – Murino – Ladozhsky – Pavlovsk/ Kolpino. All this can be realized by building platforms at an average distance of 2 km from each other, thereby forming the framework of the city's transport system and from which it is already possible to create a lower-level system. This will create new connections in the city, reduce the load on the subway, as well as reduce the number of suburban routes leading from St. Petersburg to satellite cities.

The estimated location of the traffic lines along the railway tracks. These proposals take into account the immediate plans for the construction of the metro. The middle stage of transport is trams and trolleybuses The next, lower stage will be trams or especially large trolleybuses. According to these rules and taking into account passenger traffic (there are 6-7 frequent routes on these sections) There are not enough dedicated tram lines or trolleybus trains. - Leninsky Prospekt from the shore of the Gulf of Finland, crossing the Lensky Prospekt metro station, Leninsky Prospekt railway station, Moskovskaya metro station, then along Tipanova Street, through the Prospekt Slava railway platform, through the Prospekt Slava metro station, through the Sortirovochnaya railway station, through the Lomonosovskaya metro station, and along Narodnaya Street to the Neva railway station and possibly its bring it to the Dybenko Street metro station and, in theory, through Bolshevikov Avenue up to Shafirovsky Avenue or Ladozhsky railway station. Thus, this tram will connect 5 metro lines, 4 city train lines, and many residential and industrial areas.

There are also two tram routes 36 and 41, which are a continuation of each other. They could be combined and possibly launched along the avenue of strikes. There are no ring lines in St. Petersburg, and for this reason, metro stations in the center of St. Petersburg are overloaded. Routes running perpendicular through the entire city along the main highways could solve this problem. Such routes should run in dedicated lanes, and their rolling stock should be of particularly large capacity. As an example, you can see the Leninsky Prospekt chord from the shore of the Gulf of Finland, crossing the Lensky Prospekt metro station, Leninsky Prospekt railway station, Moskovskaya metro station, then along Tipanova Street, through the Prospekt Slava railway platform, through Prospekt Slava metro station, through Sortirovochnaya railway station, through Lomonosovskaya metro station, and along Narodnaya Street. to the Neva railway station and it is possible to bring it to the Dybenko Street metro station and, in theory, through Bolshevikov Avenue up to Shafirovsky Avenue or Ladozhsky railway station. At the moment, there is a bus route 239 that almost completely fulfills this role. But it has too long an interval for mainline transport (13-20 minutes instead of 5-10) and a small capacity of rolling stock (large class buses that can accommodate 90-100 people according to standards). On such a route, given the absence of narrow streets on this route, it would be possible to launch trolleybuses that can accommodate 190-220 people, as was done in Prague (Godnev, 2020).

Let's see if such a system will be effective.

The existing bus 239 is duplicated by other routes for more than 2/3 of their length.:

- Bus 239 20 large class cars 100 people interval 15 minutes (400 people per hour);
- Trolleybus 32 16 large class 100 people interval 9 minutes (650 people per hour);
- Trolleybus 35 16 large class 100 people 15 minutes (400 people per hour);
- Trolleybus 45 19 large class 100 people 12 minutes (500 people per hour);
- Trolleybus 279 large class 100 people 25 minutes (220 people per hour);
- Trolleybus 29 13 large class 100 people 15 minutes (400 people per hour);
- Bus 114 25 large class 100 people 15 minutes (400 people per hour);
- Bus 11 22 large class 100 people 10 minutes (600 people per hour);
- Bus 140 12 large class 100 people 15 minutes (400 people per hour). (URL: https://www.24tr.ru/spb/suburban/)

The auhtors will divide the highway into sections to more accurately determine passenger traffic in a particular location.

1 - Leninsky Prospekt from the bay to Kubinskaya Street

- 239 (400 people);
- 35 (400 people);
- 32 (650 people).

A total of 1,450 people.

2 - Leninsky from Kubinskaya and Tipanova

- 239 (400 people);
- 114 (400 people);
- 35 (400 people);
- 45 (500 people);
- 27 (220 people);
- 29 (600 people per hour);
- 11 (600 people per hour)

A total of 3120 people.

3 - Slava Avenue

- 239 (400 people);
- 114 (400 people);
- 35 (400 people);
- 27 (220 people);
- 29 (600 people per hour);
- 11 (600 people per hour).

A total of 2,620 people.

4 - Sorting bridge, Ivanovskaya and Narodnaya Streets

- 239 (400 people);
- 27 (220 people);
- 140 (400 people).

A total of 1,020

It is also worth considering in these calculations that they partially intersect on the second section, and there are not so many other routes that duplicate them. At the same time, there are many routes on the 4th section that go further, but at the same time it is completely duplicated, that is, passenger traffic differs by a large margin on it.

The situation is similar on the 1st section, but there is a similar number of routes lower, so the value does not differ so much. The approximate passenger traffic data, taking into account these parameters, will be:

- 1800 per hour;
- 3,200 per hour;

- 2,800 per hour;
- 1,900 per hour.

Which tells us that the minimum value per hour when using such rolling stock will be 15 units. Therefore, the interval will be 4 minutes. In order to increase it, but to maintain the volume of traffic, and given the increased passenger traffic, since there will be a through passage through the city on one route, several methods can be resorted to.

Laying dedicated tram tracks and creating full-fledged mainline transport, and increasing the speed of movement on it. This option is more expensive, but in the future, given the increased speed of movement, it can replace the southern half–ring of the metro as a cheaper alternative, and in combination with running trains along the Ladozhskaya - Avtovo tracks, it will completely eliminate the need for this, provided convenient transfers are arranged. A similar thing can be done in the north of the city. Alternatively, it's possible to duplicate it on the busiest sections by building another route that will then leave along other streets. For example, a route can be away from Moskovskaya Street towards Borovaya, and from Sofiyskaya Street towards Obukhovo. This idea is also just an assumption. To determine more accurate routes, of course, it is necessary to obtain data on the expected passenger traffic.

The lower level of transport is buses

Buses in the city should not be one of the main types of ground-based street transport in such a large city as St. Petersburg. Its task is to transport passengers to areas with low passenger traffic. This task is assigned to him for a number of reasons.

- Capacity – buses have a much smaller capacity than other modes of transport and it is impossible to increase it.
- Economically, it is much more expensive to run multiple buses with a short interval than a slightly rarer but more spacious tram or trolleybus.
- Service life – despite the fact that buses are cheaper than all other types of rolling stock, they are the most expensive in the long run.

Service life and price-to-year ratio per vehicle unit:

- The bus is 5 years old;
- Trolleybus is 15 years old;
- The tram is 25 years old;
- The metro is 31 years old;
- City train ≈ 40 years old (depends on the specific model).

Price per unit:

- A bus with a particularly large capacity of 30 million (100 passengers);
- Trolleybus with a particularly large capacity of 45 million (120 passengers);
- Three-section tram 160 million (260 passengers);
- Metro 8 cars, 640 million (3,500 passengers);
- Urban electric train with 5 wagons of 600 million (1000 passengers).

Price per year of operation:

- Bus 30/5= 6 million per year;
- Trolleybus 45/15 =3 million per year;
- Tram 160/25=6.4 million per year;
- Metro 640/31=20.6 million per year ;
- City train 600/40 = 15 million per year.

Considering the maximum possible passenger traffic, the following price is calculated for 1 passenger seat per year:

- Bus 6000000/100=60000;
  Trolleybus 3000000/120=25000;
- Tram 6400000/260=24600;
- Metro 20600000/3500= 5880;
- City train 15000000/10000=15000.

As can be seen from rough calculations above, the cost of a bus as a rolling stock is the highest of all. If passenger traffic in some area is too high for a bus, then it is better to use other types of transport. Trolleybuses are good to use only in the case of an already established passenger traffic, since this type of transport requires additional infrastructure. It becomes clear that buses can only be used as transport for transporting passengers from mainline transport to residential areas with low population density. Therefore, the point of directness for bus routes during the construction of the system may be relevant only in places with low passenger traffic and where this type of transport is the main one, or in sections between main modes of transport. Throughout the entire route, it can form a semicircle, if necessary. In conclusion, we can put it briefly – the bus is an auxiliary mode of transport in the city, and it cannot become the main way to travel in a large city.

Mile Transportation and Its Infrastructure

A significant part of all such public transport systems is a well-developed last-mile transportation system. These are bicycles and scooters. This includes their rental system, as well as places for their travel, that is, bike paths. They are the ones that make it possible to greatly increase the girth of all public transport. And raise it from a comfortable 200-400 meters from the house to 1 km.

In St. Petersburg, on the one hand, it is being developed by creating new bike paths, and even making convenient crossings across the streets, with a separate descent to the level of the roadway and a traffic light, such as on Lunacharsky Avenue, but at the same time bans the use of scooters in some areas of the city, instead of creating infrastructure for them. Image 39 shows the presence of a special traffic light for cyclists, and a separate bicycle crossing. It's not just that people liked this type of transport. Living in Murino, in the summer, I observe from my own experience how many people leave the subway and use a scooter or an electric bike to get home 1– 1.5 kilometers from the station, and sometimes 2 kilometers away. And this significantly reduces the burden on public transport. In the cold season, due to the unpreparedness and complete unsuitability of the infrastructure for this transport, whole queues for buses form at public transport stops to get to the subway. And this greatly affects the comfort of passengers and discourages them from using public transport. And for bike paths, Russian regulations have all the necessary parameters, which allows them to be organized on city streets without the need for separate approvals.

CREATION OF A PUBLIC TRANSPORT SYSTEM AND ITS JUSTIFICATION

At the moment, there are many duplicate routes in St. Petersburg, and some of them duplicate mainline transport, such as the 50 bus duplicates the 2nd metro line from Sennaya to Kupchino. At the moment, the clock movement of suburban trains is being created in St. Petersburg.

On the one hand, this is correct, but all this modernization is being carried out with a number of grave mistakes.:

- Large difference in intervals during the day (10 minutes during rush hours and 30-40 minutes the rest of the time);
- Breaks on weekdays;
- City train tickets are not included in the general ticket system of the city;
- The rest of the transport system has not changed in any way.

Paragraph 1 and paragraph 2 are closely interrelated and have similar consequences. On the one hand, this helps to relieve the load on the road system, but not to the extent that it could if trains moved at approximately the same intervals throughout the day. At the current moment, the intervals vary from 10 minutes to 40, and also in the middle of the day on weekdays there is a gap of 2 hours, which leads to reduced passenger traffic and for the most part only connects satellite cities with St. Petersburg, although electric trains have the potential to become a full-fledged intercity transport, as it has become in Moscow, for example. in recent times, and what has been presented in this paper.

Since commuter train tickets were not included in the general ticket system of the city and the price of tickets still depends on the distance, many people who need to get closer to the city center still transfer to the subway, thereby increasing the load on the metro system and not using rail transport to the fullest. The fact that the rest of the public transport system has not changed in any way means that no more convenient transfers have been made, and the route of land transport routes has not been changed. As an example, person can cite the Prospekt Slava railway platform.

Despite the fact that now a train passes through this station every 10-15 minutes during rush hours, the nearest public transport stop is located 250 meters from the platform from Vitebsk Avenue. Although it was possible to change the location of the bus stop by moving it closer to the underpass. Also, on the opposite side, the stop is made in the pocket, which in this situation contradicts even the current GOST, since the pocket according to it should be made only when the road is heavily loaded with an indicator above 0.6. Fines and car evacuation do not stop drivers from violating and they park their cars in the bus pocket, thereby forcing the bus stops further from the edge of the road. Belgradskaya Street, in turn, has low traffic. In an ideal situation, an antikarman would be required, as previously described. In this case, there would be more parking spaces, and the bus could always drive closer to the edge of the road, increasing convenience for passengers. Moreover, the stop would be located away from the pedestrian crossing, shortening the transfer path.

There are also still not enough dedicated lanes for public transport in St. Petersburg, but they greatly increase the capacity of the street in people. So, according to the data, the allocated lanes for public transport allow to increase the speed of buses, trams and trolleybuses by 15-30%. For example, after the introduction of dedicated lanes in Moscow on the Prospekt Andropova — Kolomenskoye section, the speed of public transport increased to 43% (Interfax, 2011).

This also correlates with my personal experience working at Domtransauto. Route 249, despite the fact that it passes through the central districts, is extremely rarely more than 20 minutes late when returning from a 3-hour circle, and most often these delays are associated not so much with traffic congestion as with poor street cleaning, or traffic accidents that have blocked traffic, completely or partially. The image shows which part of the route runs along the designated lanes and which does not, as well as the most frequent traffic jams. Green – places where there is a dedicated lane, red – where there is no dedicated lane, black – places where traffic jams occur most often. The reverse example is route 160.

There are no dedicated lanes throughout it, although in some places it is necessary. First of all, this concerns the Boulevard of Innovators. In my opinion, this street should be closed to private transport from Veterans Avenue to Kuzmin Submariner Street, since the Prospekt Veteranov metro station is located on it, and in general this place is an important interchange hub. In addition, this closure will increase the speed of personal transport along Dachny Prospekt, as there will be no left turns, which means it will also reduce the number of traffic jams at the exit and exit from the ring road. Marshal Zhukov Avenue and its arrivals are also problematic. The creation of dedicated lanes on adjacent streets may be unnecessary, since they have only two lanes of traffic and congestion on them does not occur on a regular basis, and they are not so serious. There are 3 lanes of traffic on the avenue itself, and congestion on the avenue itself is quite serious and occurs on a regular basis, especially in winter.

CREATING GROUND TRANSPORT TRAFFIC PATTERNS

In this research, the authors propose to consider the Pushkin public transport system and its connections with the rest of the districts, using the derived rules. Pushkin has transport links with Krasny Selo, Kolpino, Kommunary, as well as with Kupchino – Zvezdnaya, Moskovskaya and a number of other small towns. Pushkin also has a railway connecting it with Kupchino and Vitebsk railway station. Since authors are currently considering a small town and its connection to other areas, there is no reason to consider infrastructure such as trams and trolleybuses, since these projects must have a constant passenger flow. This can be implemented only with the stability of large passenger flows on these routes.

A total of 264 units of rolling stock are used on these routes.

According to this table, it turns out that on all routes, in total, all drivers work 1909 hours per day.

According to this table, it turns out that on all routes, in total, all drivers work 1,423 hours per day, which is less than 1,909 by 486 hours, or 519395 hours per year instead of 696785 per year. The scheme was created stemmed from the following logic: an intersection should be created inside the city and a transfer from the maximum possible number of routes is possible. Thus, reducing the necessary number of transfers to get to any point of the city or surrounding towns. Bus routes leading from Pushkin to Kupchino station were also removed due to their uselessness, due to the duplication of the clocked railway network. Duplication of the routes themselves has been minimized to increase the coverage of the transport network. At the same time, a large distance to the nearest public transport stop was not allowed. The intervals were designed taking into account possible passenger flows, but with minimal time difference from each other.

DISCUSSION

Creating a Regular Transportation Route Connecting 3 Municipalities of St. Petersburg's Krasnogvardeysky District

The transport system of such a megalopolis as St. Petersburg is quite complex and, often, incomprehensible due to the general framework of streets on which any transport system is formed. However, like any system, the transport system has its own problems. For example, the biggest problem is the small coverage of residential areas, since social transport in St. Petersburg covers only 60% of residential areas, and 40% of residents use transfers or taxi. (Vuchic V.R. Urban Transit Systems and Technology)

However, no urban logistics system can fully cover all residential buildings, because with so many routes, the streets of the city will lose the required capacity due to the large number of buses, passenger traffic will be small (residents will be evenly distributed on all routes), and the economic efficiency that any carrier strives for will drop sharply. Naturally, any passenger transportation is economically unprofitable, since the cost of their maintenance exceeds the income from passenger transportation – otherwise, a trip on urban public transport will cost close to a taxi ride, without any benefits and discounts! Accordingly, the transport department faces the task: how to create the most efficient and cost-effective route system at the same time? What needs to be considered and what needs to be relied on to create such a system?

1. The city's street and road network.

It should serve as the main support for building any routes. Particular significance is the linking of the road to certain objects that can be reached by moving along the road and how fast; the width of the road (the number of lanes will determine the capacity – and from here the possibility of creating a dedicated lane or tram line is determined), as well as safety (the presence of traffic congestion along the road, places of concentration of accidents, the number of traffic lights, pedestrian crossings). Each of these indicators either allows you to call the road an arterial highway (that is, wide, free, safe), or secondary (narrow, with low capacity, not safe enough). Accordingly, it is immediately clear that the arterial highway is more profitable: vehicles will travel faster, and perhaps even separately from the general flow of cars, since the road can be equipped with dedicated lanes.

The movement of vehicles along arterial highways alone is not convenient for passengers and is not beneficial for carriers, since the majority of residential areas are equipped mainly with secondary roads, because arterial roads require more space, and exhaust gases emitted by a large flow of cars have a bad effect on the quality of life in those residential areas. That is why the route network is created in this way: there are many routes along arterial highways, but they all diverge along roads of secondary importance. Therefore, there are only 2-3 routes to the residential areas themselves, and at metro stations or in the city center their number reaches 15-20. But even with this orientation, the route network cannot be considered effective. (http://elibrary.ru/item.asp?id=47486832)

2. Rapid transit systems (metro, railway, and rapid tram stations)

The logic of travel for most city residents is clear: passengers often cannot directly reach their planned destinations. To do this, they need to use a fast and comprehensive system (rapid tram, subway). Therefore, the main passenger traffic moves from home to the metro stations, and then they take the subway to the object they need, to which a public transport route is sometimes also needed from the metro station. This means that it makes no sense to take every residential block to the city center (and most residents are heading there). It is enough just to run 1-2 routes to the city center along the arterial highway. They will drive longer, as there is a large traffic flow from all the arteries on the way to the city center, and traffic jams occur on the approaches to the city center. Therefore, few passengers will choose this type of transport. Most of them will prefer to get to the center by high-speed transport. Therefore, each neighborhood must be connected to the metro station by a land transport route. This can be done using a single route that passes through all neighborhoods, but this option is beneficial only for small neighborhoods. And for large residential complexes, it is more humane to make several routes, since all passengers will not be able to accommodate in a crowded bus cabin. (http://elibrary.ru/item.asp?id=47486832)

Now let's move from theory to practice and consider the planned route. (see Figure 1).

Figure 1. Projected route

The first issue is, as always, traffic safety. The authors used the resource dtp-stat.com, which assisted the authors to find statistics on accidents involving public transport in St. Petersburg for 4 years, and uploaded it to the map. By superimposing the projected route layer on the accident layer, you can see how safe my route will be. If the planned route does not avoid all traffic congestion, which is impossible in such busy districts as Bolshaya Okhta or Porokhovye, then at least it does not fall into the most unsafe places according to statistics (see Figure. 2).

Figure 2. Places of cars' congestion and accidents' concentrations

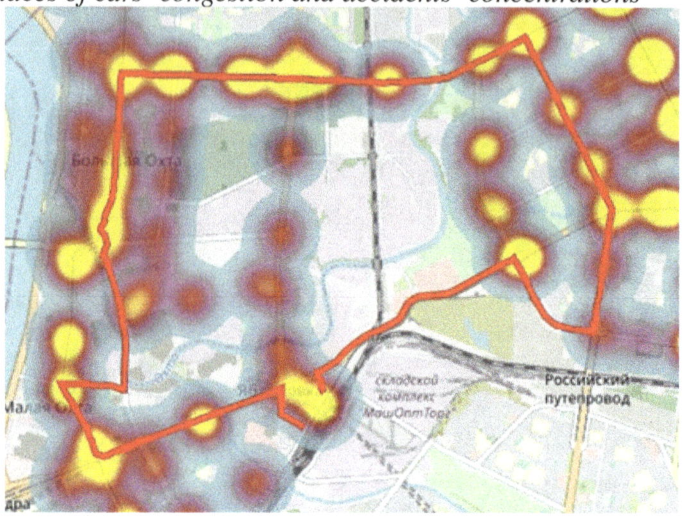

Residential areas and the expected number of passengers were considered. The work was carried out in the ArcGIS program using the overlay method (layering on the map). In other words, layers of stops, routes, traffic congestion and accident concentrations are added to a regular map, and the route data is ready: you can click on any stop or street and a sign with data for each of them will appear.

There is also a very important and necessary function: to designate buffer zones around stops. Their radius is set at 400 meters, as the SNiP says: "6.29. The distance of pedestrian approaches to the nearest public transport stop should be no more than 500 m; the specified distance should be reduced in climatic subdistricts IA, IB, IG and IIA to 300 m, and in climatic subdistrict ID and climatic region IV to 400 m. " St. Petersburg is located in the IV climatic region. When all the buffer zones were highlighted, those houses that did not get into them remained. Such areas are called "dead corners". There is an acute shortage of public transport, and residential buildings are present. Therefore, my route also covered two dead corners: Respublikanskaya St. – Shahumyan Ave. (construction of new residential complexes), and Peredovikov St. (housing stock since 1986; bus route 21 previously passed, but it did not completely cover the dead corner).

Using the ArcGIS program, the data is calculated and the results for each stop are displayed for certain time intervals: $6^{30} – 10^{00}$, $10^{00} – 16^{00}$, $16^{00} – 20^{00}$, $20^{00} – 22^{30}$, и $22^{30} – 1^{00}$. Stops have been identified, a table has been compiled, and this data has been entered into it. The expected number of passengers at the missing (new) stops was calculated using a weighted average of neighboring.

The stops of the route were investigated in more detail. According to Yandex research data, based on the popularity of ground transport stops, it was predicted which stops are most in demand, and this fact was taken into account when designing the route (see Figure 3).

Figure 3. The popularity of public transport stops (approximate time is 18:00)

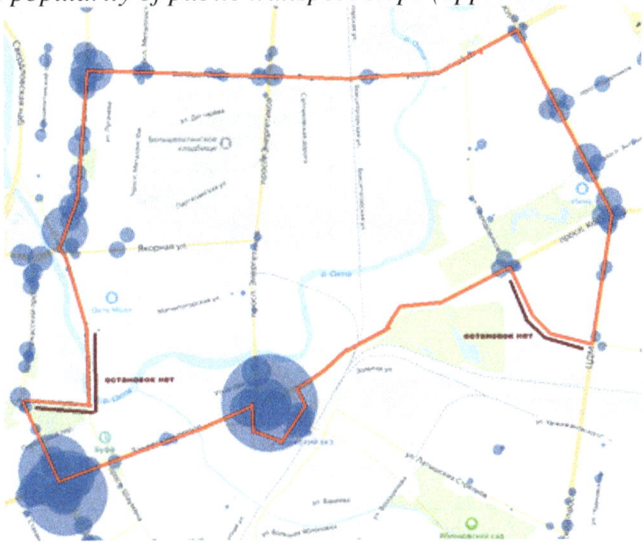

Next, it was necessary to calculate the remaining data to create a route (see Table 1).

Table 1. Calculation table for the projected route

Average travel time on the route, *I*, minutes Determined by the survey data (at what time passengers use the airport more often)	$6^{30} – 10^{00}$ $10^{00} – 16^{00}$ $16^{00} – 20^{00}$ $20^{00} – 22^{30}$ $22^{30} – 1^{00}$ 10 15 15 10 15
Number of trips, N_t Calculated by next formula: $N_t = \frac{t}{I}$	$6^{30} – 10^{00}$ $10^{00} – 16^{00}$ $16^{00} – 20^{00}$ $20^{00} – 22^{30}$ $22^{30} – 1^{00}$ 33 24 16 23 10
Total number of trips, $N_{общ}$ Calculated by next formula: $N_{общ} = N_{6\text{-}30-10\text{-}00} + N_{10\text{-}00-16\text{-}00} + N_{16\text{-}00-20\text{-}00} + N_{20\text{-}00-22\text{-}30} + N_{22\text{-}30-01\text{-}00}$	106
Passenger turnover, average, per day, $П_д$, passengers	39 253
Passenger turnover, average, for trip, $П_р$, passengers	$\frac{39\ 253}{106} = 370$
Passenger turnover, average, per turnover, $П_о$, passengers $П_о = П_р \times 2$	$370 \times 2 = 740$

continued on following page

Table 1. Continued

Revenue per trip, average, R_p, Calculated by next formula $R_p = П_p \times R_{ЕКП}$, where $R_{ЕКП}$ – the price of paying for trip on a Single St. Petersburg Card, fixed for ULPT	$370 \times 40 = 14\ 800$
Revenue per turnover, average, R_o, Calculated by next formula $R_o = П_o \times R_{ЕКП}$, where $R_{ЕКП}$ – the price of paying for trip on a Single St. Petersburg Card, fixed for ULPT	$740 \times 40 = 29\ 600$
Velocity, V_{cp} Calculated by next formula: $V_{cp} = \frac{Vmax + Vmin}{2}$	<math xmlns="http://www.w3.org/1998/Math/MathML" xmlns:tps="http://www.typefi.com/ContentXML" xmlns:xd="http://www.pnp-software.com/XSLTdoc" xmlns:xsi="http://www.w3.org/2001/XMLSchema-instance" id="m5"> <mfrac> <mrow> <mn>60</mn> <mo>+</mo> <mn>0</mn> </mrow> <mrow> <mn>2</mn> </mrow> </mfrac> </math> = 30
Route length, $L_{общ}$, km	13
Bus journey time on the route, $t_{p(общ)}$, minutes Calculated by next formula:	47
The turnaround time of the bus on the route, t_o, minutes	$(47\times 2) + 15 = 109$
Passenger capacity of buses, (amount of people),	Class "B" (Large), up to 90 seats Standing places: 56 Seating places: 37 (33+4 folding) Places for disabled: 1 (33 seating places)
Number of buses on the route, V_A Calculated:	<math xmlns="http://www.w3.org/1998/Math/MathML" xmlns:tps="http://www.typefi.com/ContentXML" xmlns:xd="http://www.pnp-software.com/XSLTdoc" xmlns:xsi="http://www.w3.org/2001/XMLSchema-instance" id="m6"> <mfrac> <mrow> <mi>X</mi> </mrow> <mrow> <mn>47</mn> </mrow> </mfrac> </math> = $\frac{90}{370}$ => $\frac{47 \times 90}{370}$ = 11,4 ≈ 12

continued on following page

Table 1. Continued

Vehicle type selection It is carried out by the availability of charging infrastructure. If there is an $L_{KC} > L_{AX}$ on the route, it is more advisable to use Electric bus with charging on the move	7, 62 km of the route passes under the contact network (L_{KC}). 5, 38 km of the route passes without a contact network (L_{AX}). 7,68 > 5,38 => use Electric bus with charging on the move
Type of vehicle used on the route	Electric bus with charging on the move
Type of vehicle model	KAMAZ-62825
Vehicle's photo Figure. 4 Electric bus with charging on the move KAMAZ-62825 (on the right side) [4] The following 7 parameters are taken from the RE of this vehicle [C. 14]	
Accumulator type	LTO
Rated voltage, B	440
Capacitance, A×h	90
Energy intensity, kWh	40
Full charge time, minutes	20
Autonomous distance, км	20 и более
Equipment	IMC – In Motion Charging Recovery system
The cost of all vehicles used on the route, $C_{\Sigma TC}$ Calculated by next formula: $C_{\Sigma TC} = C_{TC} \times V_A$	20 000 000 (C_{TC}) × 12 (V_A) = 240 000 000
DTI (charging stations, traction substations)	- (not required)

(UN-Habitat, 2020).

In those three municipalities, the total number of routes exceeds 37. Of these, 25 duplicate the planned route in some way (along at least one of the streets). The projected route is mostly duplicated by the 174 bus route. Its route coincides with the projected one by 20.6% (see Figure 4). Duplication is taking place along the following sections (see Table 2).

Table 2. Table of duplicated plots

Duplicated Section	Length, m
Industrial Ave	1019
Sredneokhtinsky Ave	1126
Novocherkassky Ave	523
Zanevskiy Ave	163

continued on following page

Table 2. Continued

Duplicated Section	Length, m
Total length of duplicated sections	2682
Total length of the bus route 174	13785
Calculating the duplication percentage	20,6%

Figure 4. Projected route (red) and bus route 174 (orange)

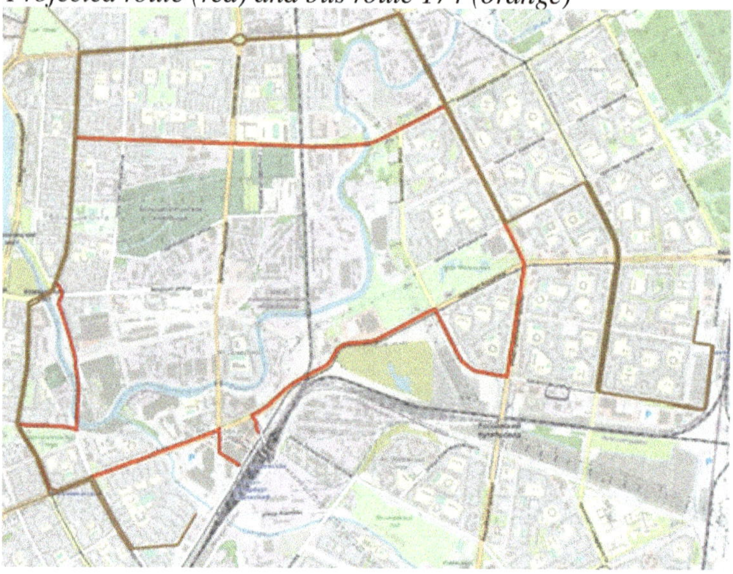

A number of pros of the projected route in comparison with the 174 bus route are highlighted. These include:

1) Direct connection of residential areas with metro stations along the main arteries of the Krasnogvardeysky district (Zanevsky Ave., Novocherkassky Ave., Sredneokhtinsky Ave., Bolshaya Porokhovskaya St., Industrialny Ave., Kosygina Ave.);
2) The projected route practically excludes places of concentration of accidents and places of congestion of cars;
3) Unlike the bus route, 174 passes through 2 metro stations, which means it is more in demand by passengers than the last one;
4) Unlike bus route 174, the projected route is serviced by the number of units necessary for comfortable and safe transportation on this route. On route 174,

only half of the number of units needed to ensure the speed and comfort of transportation goes on the line.

A number of disadvantages of the projected route have been identified, but they also cannot be attributed to the advantages of the 174 bus route:

1) To cover all residential areas, the route does not connect a specific block directly to the metro station (there are existing routes for this), but only passing through all the blocks.
2) There was a 20% alternate route (bus 174)
3) Putting the maximum number of vehicles on the route may be economically unprofitable, even if the profit from the flight is high. Below is a less detailed and more specific map of routes that duplicate the one being designed for each of the streets (see Fig. 5). (http://elibrary.ru/item.asp?id=47486832)

Figure 5. A map of routes that duplicate the planned one along at least one street

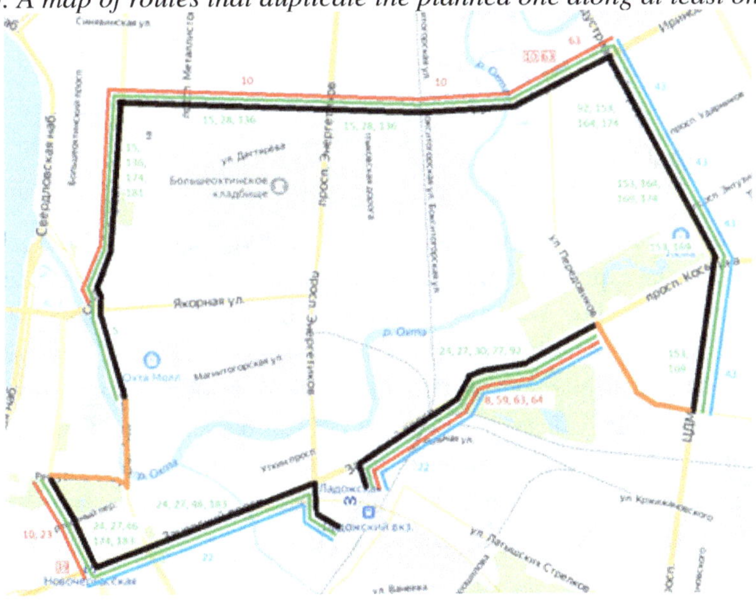

The route was visualized in the ArcScene program in order to see which residential buildings the route covers. To do this, I chose the 3D style and stretched the houses. As a result, the route covers almost all the houses along the way.

Using simulation modeling in the AnyLogic program, the effectiveness of the projected route was tested, the route was fully visualized and various situations were modeled.

So, the result of the study is the creation of a regular transportation route that is optimal in the following parameters:

- speed of movement (direct connection of residential areas along arterial highways);
- safety (avoiding places of concentration of accidents and places of congestion of cars);
- passenger comfort (bringing the estimated number of vehicles onto the line).

It should be added that the creation of a regular transportation route cannot do without exploring the territory for transport accessibility using various programs, namely ArcGIS, QGis and others. Such a study can substantiate the optimality and logic of the route, as well as its relevance for the territory for which it is being drawn up. An equally important task in the process of creating a route is to visualize the route using programs such as ArcScene and AnyLogic. They can visually show the effectiveness of the route.

CONCLUSION

In this scientific research, various public transport systems of European cities were presented, the experience of their implementation in Moscow was considered, as well as examples of how such a system could function in St. Petersburg. The main similarities and differences between the transport systems of various megacities are highlighted. The main parameters that should be considered when creating new urban transport and logistics systems in the field of passenger transportation have been identified. There are 2 examples of developing mixed routes.

These aspects have been taken in a highly condensed form, but even so it is clear that this system has a number of advantages:

- It is cheaper to service due to the avoidance of duplication of expenses for performing the same tasks
- It is more convenient for passengers, as it reduces the number of transfers and the time spent on them due to longer public transport routes and network arrangements so that different routes intersect with each other
- More convenient and intuitive navigation in the city by reducing the number of routes.

It should be borne in mind that all of the above works only in combination. To implement only a part of these innovations, ignoring the rest may either not work enough or make the city's transport system even worse. It should be borne in mind that such modernization has been a matter of many years, as shown by the experience of Vienna, where the system was upgraded from a state in which residents did not want to use public transport to a system where public transport became the main way to move inside the city, thereby accelerating the movement of residents inside the city for more than half a century, since the late 60s and until the early 2010s. All this time acting according to the same plan, only correcting it in the process and correcting the shortcomings.

REFERENCES

Berliner Verkehrsbetriebe. (n.d.). Connection search. https://www.bvg.de/en/connections/connection-search

Dopravní podnik hlavního města Prahy. (n.d.). Ceník jízdného [Fare prices]. https://www.dpp.cz/jizdne/cenik-jizdneho

Duranton, G., & Turner, M. A. (2011). The fundamental law of road congestion: Evidence from US cities. *The American Economic Review, 101*(6), 2616–2652. DOI: 10.1257/aer.101.6.2616

Godnev, A. (2020). Intercepted parking as a method of combating transport convenience of the megapolis. salogistics.ru. http://www.salogistics.ru/magazine/26/15_godnev_115-121.pdf

Interfax. (2011, December 2). В Москве утверждены новые тарифы на проезд в общественном транспорте [New public transport fares approved in Moscow]. https://www.interfax.ru/russia/217432

JLL. (2017). *Residential city profile: Berlin*. https://jll-global-gdim-res.cloudinary.com/image/upload/v1524021097/olmwebcontents/ip/prod/mmlimougy5wnrbmzhrom.pdf

Knoflacher, H. (2017). Verkehr ist kein Schicksal: Der öffentliche Verkehr in Wien [Traffic is not fate: Public transport in Vienna]. TU Wien. https://www.tuwien.at/index.php?eID=dumpFile&f=138722&t=f&token=b57c5dd11329b20318241ed0048d2c66474a7018

Live Streets. (2013, October). Пересадки или прямые маршруты? [Transfers or direct routes?]. http://www.livestreets.ru/2013/10/peresadki-ili-pryamye-marshruty/

Prillinger, H. (n.d.). Tarife [Fares]. University of Vienna. https://homepage.univie.ac.at/horst.prillinger/ubahn/deutsch/tarife.html

Rambler. (2024, August 4). Пассажиропоток наземного транспорта Москвы вырос на 6 млн человек в марте [Passenger traffic of Moscow's ground transport increased by 6 million people in March]. https://news.rambler.ru/moscow_city/52541549passazhiropotok-nazemnogo-transporta-moskvy-vyros-na-6-mln-chelovek-v-marte/

UN-Habitat. (2020). *International guidelines on urban and territorial planning*. https://unhabitat.org/sites/default/files/2020/04/ig-utp_russian.pdf

Chapter 7
Municipal Policies for Autonomous Transportation in Critical Sectors of Food and Clothing in Russia: A SWOT Review

Galina Vladimirovna Astratova
https://orcid.org/0000-0002-3579-4440
Ural Federal University Named After the First President of Russia B.N. Yeltsin, Yekaterinburg, Russia

Chigozirim Ndubuisi Onwusiribe
https://orcid.org/0000-0002-7740-5458
Ural Federal University, Russia

Callistus Destiny Odeh
https://orcid.org/0009-0004-5441-9023
Ural Federal University Named After the First President of Russia B.N. Yeltsin, Yekaterinburg, Russia

ABSTRACT

The study analyzed municipal policies for autonomous transportation (AT) in Russia's food and clothing sectors using a SWOT analysis framework. Digitalization reshapes urban infrastructure, presenting AT as a means to enhance supply chain efficiency, reduce congestion, and strengthen economic resilience. However, challenges like regulatory ambiguity, infrastructure deficits, and regional disparities remain. The

DOI: 10.4018/979-8-3373-6335-6.ch007

Copyright © 2026, IGI Global Scientific Publishing. Copying or distributing in print or electronic forms without written permission of IGI Global Scientific Publishing is prohibited.

study identifies strengths in technological innovation and economic capacity, weaknesses in inconsistent municipal planning, opportunities for improved food security in remote areas, and threats of environmental harm. Key findings stress the need for municipal-private partnerships, targeted policy changes, and adaptable governance structures. Recommendations include integrating AT into multimodal networks, promoting Arctic logistics investment, and establishing a formal municipal social logistics solution for food and clothing, contributing to evidence-based practices for improving transportation policies in developing economies.

1. INTRODUCTION

The convergence of digital transformation imperatives and urbanization challenges has positioned autonomous transportation (AT) as a critical component of 21st-century municipal policy frameworks, fundamentally reshaping how cities approach mobility, logistics, and service delivery (Nikitas et al., 2017; Zhang et al., 2020). In the context of globalization and digitalization of the economy (Astratova et al., 2024), autonomous transport systems that utilize artificial intelligence, optical sensors, and radar for driverless management represent a paradigmatic shift in transportation governance and urban policy implementation (Lazutkina et al., 2019).

Russia's unique geographical vastness, diverse municipal capacities, and federal governance structure create distinctive implementation challenges and opportunities for AT integration, particularly in critical consumer sectors such as food and clothing (F&C) that are essential for basic human needs fulfillment and economic stabilization through workforce solidification and consumer spending (Dementiev & Loboyko, 2014; Petrov & Kolesnikov, 2021). The Russian Transport Strategy explicitly recognizes the implementation of digital technologies and AT development as paramount tasks, positioning municipal transport policy as a crucial supporting mechanism for food and clothing distribution networks.

Research Context and Theoretical Significance

This study addresses a critical gap in existing literature by providing the first comprehensive SWOT analysis of Russian municipal AT policies specifically focused on critical consumer sectors, incorporating stakeholder perspectives, financial modeling, and international benchmarking. The theoretical significance lies in its integration of New Institutional Economics (NIE) and Multi-Level Perspective (MLP) frameworks to understand policy implementation dynamics across Russia's multi-tiered governance structure, while the practical importance stems from growing recognition that effective AT deployment requires coordinated municipal policies addressing both technological capabilities and socio-economic integration challenges (Mladenović et al., 2019; Soteropoulos et al., 2019).

The food and clothing sectors, representing approximately 40% of urban freight movements and directly impacting citizen welfare, serve as critical test cases for AT policy effectiveness (Rosstat, 2023). These sectors extend beyond economic considerations to encompass cultural identity, social cohesion, community well-being, and innovative approaches to economic sustainability (Dementiev, 2016). Recent developments in Moscow's autonomous bus program and Kazan's smart logistics initiatives provide concrete examples of municipal innovation, while persistent challenges in smaller cities highlight the need for evidence-based policy frameworks.

Evolution of Municipal Transport Strategies

Municipal transport strategies in Russia have undergone dramatic transformation since the 1990s, transitioning from state-owned approaches to frameworks incorporating private operators through market initiatives and subsequent reforms that fundamentally altered the transport sector landscape (Ryzhkov & Sarzhan, 2020; Astratova et al., 2021). This evolution has created a complex policy environment characterized by high degrees of disparity in transport infrastructure and access between regions, with many areas lacking digital infrastructure, modern road networks, and requiring regionally targeted or pilot solutions rather than blanket approaches (Zamyatina et al., 2020).

The current Russian municipal governance framework operates within a decentralized structure where local authorities possess varying degrees of autonomy in managing public services, including transportation, allowing for regional differences in governance structures, fiscal capacities, and service delivery methods (Dementiev, 2016; Komarov et al., 2019). This framework is further complicated by vast territorial coverage featuring both densely populated urban centers and sparsely populated regions, necessitating diverse transportation solutions adapted to local contexts (Zamyatina et al., 2020; Zavyalov, 2021).

2. THEORETICAL FRAMEWORK AND LITERATURE REVIEW

New Institutional Economics and Municipal Policy Implementation

Drawing on North's (1990) seminal institutional analysis framework, this study examines how formal and informal institutions shape AT policy implementation across Russian municipalities. The NIE perspective emphasizes transaction costs, institutional complementarity, and path dependency in explaining policy outcomes, providing analytical tools for understanding variations in municipal AT adoption rates and implementation effectiveness (Williamson, 2000).Recent empirical studies demonstrate significant institutional barriers to technology adoption in Russian municipalities, including regulatory ambiguity, capacity constraints, and coordination challenges between federal and local authorities (Golovin & Rukavishnikov, 2022; Makarova et al., 2021). These findings align with broader NIE literature on institutional barriers to innovation adoption in emerging economies, highlighting the importance of institutional design in facilitating or constraining technological transitions (Acemoglu & Robinson, 2019).

Multi-Level Perspective on Socio-Technical Transitions

The Multi-Level Perspective framework (Geels, 2002, 2011) provides analytical structure for understanding AT integration as a socio-technical transition involving complex interactions between technological innovations (niche level), existing transport systems (regime level), and broader societal trends (landscape level). This perspective proves particularly relevant for understanding how municipal policies can facilitate or constrain AT transitions within Russia's diverse institutional landscape. Application of MLP to autonomous transport transitions reveals intricate interactions between technological capabilities, regulatory frameworks, user acceptance, and infrastructure requirements (Marletto, 2019; Ryghaug & Toftaker, 2021). Russian municipalities face additional challenges related to federal-regional-local governance coordination and varying levels of technological readiness across regions, requiring nuanced policy approaches that account for local contexts while maintaining system-wide coherence.

International Policy Learning and Benchmarking

Policy transfer literature emphasizes the importance of international learning in technology adoption while highlighting adaptation challenges related to institutional contexts and local conditions (Dolowitz & Marsh, 2000; Peck & Theodore, 2015). Our analysis incorporates lessons from leading AT implementation countries to identify transferable policy innovations and adaptation requirements for Russian contexts.

Japan's Society 5.0 Initiative

Japan's comprehensive approach to autonomous system integration provides valuable lessons for coordinated policy implementation, emphasizing public-private partnerships, staged deployment, and citizen engagement that offer potential models for Russian adaptation (Cabinet Office of Japan, 2021; Shibata et al., 2022). The initiative's success in creating integrated digital ecosystems demonstrates the importance of holistic policy approaches that transcend sectoral boundaries.

Germany's Digital Transport 2025

Germany's federal-state coordination mechanisms and emphasis on intermodal integration provide relevant institutional models for Russia's multi-tiered governance structure (Federal Ministry of Transport and Digital Infrastructure, 2022; Weber et al., 2021). The German experience highlights the critical importance of institutional coordination mechanisms in managing complex technology transitions across multiple governance levels.

South Korea's K-City Program

South Korea's testbed approach and emphasis on real-world validation offer practical implementation models, particularly for smaller Russian cities seeking to develop AT capabilities (Ministry of Land, Infrastructure and Transport, 2023; Park et al., 2022). The K-City model demonstrates how controlled testing environments can accelerate technology adoption while minimizing implementation risks.

Sector-Specific Considerations: Food and Clothing Industries

The food and clothing sectors present distinct AT implementation challenges and opportunities requiring specialized policy approaches. Food sector applications necessitate consideration of cold chain management, perishability constraints, food safety regulations, and seasonal demand variations, while clothing sector applications emphasize inventory management, seasonal demand fluctuations, fashion cycle responsiveness, and supply chain flexibility (Chen & Liu, 2021; Rodriguez-Salvador et al., 2022).

Recent studies of urban freight automation highlight sector-specific requirements for successful AT integration, including specialized vehicle configurations, regulatory adaptations, supply chain coordination mechanisms, and stakeholder engagement strategies (Joerss et al., 2016; Paddeu et al., 2020). Russian applications must additionally consider geographical challenges, extreme weather conditions, seasonal variations, and diverse municipal capacity levels that create unique implementation requirements.

Research Gap and Evidence Base

Despite extensive research on transportation policy and emerging technologies, there exists a significant lack of systematic evaluations of Russian autonomous municipal transportation policies developed within structured analytical frameworks such as SWOT analysis. While scholars have examined various aspects of transportation infrastructure rationalization and policy outcomes (Dementiev & Loboyko, 2014; Komarov et al., 2019; Lazutkina et al., 2019; Zavyalov, 2021), there remains missing articulation of rigorous SWOT analysis specifically focused on autonomous municipal transportation systems in critical consumer sectors. Existing literature documents smaller-scale studies focusing on evaluating public-private partnerships (Dementiev, 2016), municipal roles in transport complex functioning (Komarov et al., 2019), sustainable mobility in megalopolises (Zavyalov, 2021), and regional development challenges (Akbulaev & Bayramli, 2020). However, the sparse utilization of structured analytical methodologies within autonomous municipal transportation systems has resulted in less effective policy development and implementation understanding (Ryzhkov & Sarzhan, 2020).

Study Objectives and Questions

1. **Conceptual Analysis**: Consider the concept of sustainable mobility in modern megacities and identify the strategic positioning of autonomous transport within comprehensive urban mobility frameworks.
2. **Policy Effectiveness Assessment**: Analyze the effectiveness of current municipal transportation policies in supporting food and clothing sector supply chains across diverse Russian urban contexts.
3. **Success Factor Identification**: Identify key success factors and implementation challenges in developing effective transportation solutions at the municipal level, incorporating stakeholder perspectives and international best practices.
4. **Impact Evaluation**: Evaluate the impact of existing policies on supply chain efficiency, urban mobility outcomes, and sector-specific performance indicators.
5. **Evidence-Based Recommendations**: Develop comprehensive, evidence-based recommendations for policy optimization using systematic SWOT analysis integrated with financial modeling and stakeholder input.

Research Questions

1. How can municipal policies for autonomous transportation in Russia's food and clothing sectors be optimized through systematic SWOT analysis to enhance supply chain efficiency, urban mobility, and economic development outcomes?

2. What are the primary strengths, weaknesses, opportunities, and threats characterizing current municipal AT policies in Russian food and clothing sectors?
3. How do institutional frameworks and governance structures influence AT policy implementation effectiveness across different municipal contexts?
4. What lessons can be drawn from international experiences in AT policy development and implementation for Russian adaptation?
5. What financial and institutional mechanisms are required to support successful AT ecosystem development in Russian municipalities?

Chapter Aim and Contribution

This chapter aims to provide a comprehensive SWOT analysis of municipal policies for autonomous transportation in critical sectors of food and clothing in Russia, contributing to the broader understanding of digital urban transportation policy in emerging economies while offering practical, evidence-based recommendations for policy optimization. The research contributes to academic literature by integrating NIE and MLP theoretical frameworks with empirical analysis of Russian municipal experiences, while providing practical value through actionable policy recommendations based on international benchmarking and stakeholder engagement.

The study's unique contribution lies in its systematic integration of theoretical analysis, empirical research, and practical policy recommendations, addressing the identified gap in comprehensive analytical frameworks for autonomous transportation policy evaluation in federal governance systems. By focusing on critical consumer sectors and incorporating diverse stakeholder perspectives, the research provides nuanced insights into the complex dynamics of technology policy implementation in diverse institutional contexts.

Literature Review

The evolving landscape of autonomous transportation (AT) systems presents significant opportunities and challenges for municipal policy development, particularly in the context of Russia's food and clothing supply chains. Contemporary scholarship has increasingly focused on the intersection of technological innovation and municipal governance, with particular attention to the implications for urban mobility and sectoral logistics efficiency.

Sustainable mobility emerges as a critical dimension in contemporary megacities, where the convergence of congestion, environmental concerns, and economic efficiency creates complex policy challenges. Freemark et al. (2019) demonstrate that while autonomous vehicles possess the potential to reduce travel times through optimized routing and reduced human error, they simultaneously risk contributing to increased total vehicle kilometers traveled, potentially exacerbating rather than alleviating urban congestion. This paradox underscores the complexity of integrating autonomous systems into existing urban transport networks without comprehensive policy frameworks that address modal integration and demand management.

The environmental implications of autonomous vehicle deployment present equally complex considerations. Research by Dawkins et al. (2023) and Ebzeev et al. (2023) indicates that while autonomous systems can reduce emissions through optimized routing and reduced idling, the potential for increased energy consumption due to higher utilization rates remains a significant concern. Bösch et al. (2018) further emphasize that the environmental benefits of autonomous systems are contingent upon their integration with sustainable energy sources and comprehensive modal planning that prioritizes shared rather than individual autonomous vehicle deployment.

Economic efficiency considerations reveal substantial potential for cost optimization through shared autonomous fleets, yet significant questions remain regarding economic accessibility for marginalized populations. Freemark et al. (2020) and van Burgsteden et al. (2024) highlight that while autonomous systems may reduce operational costs for transport providers, the distribution of these benefits across different socioeconomic groups requires careful policy attention to prevent the emergence of technology-driven transport inequalities.

The governance dimension of autonomous transport implementation presents particular challenges within Russia's federal structure. Kovin and Semenov (2022) identify significant coordination difficulties between municipal and federal authorities, noting that local governments often lack the regulatory capacity to implement comprehensive autonomous vehicle policies without federal coordination and support. This institutional fragmentation creates implementation barriers that require systematic address through coordinated policy frameworks spanning multiple government levels.

Infrastructure requirements for autonomous transport deployment represent another critical consideration, with Freemark et al. (2019) and Uldanov (2019) emphasizing that successful implementation requires substantial investments in both digital and physical infrastructure. These investments include advanced communication networks, smart traffic management systems, and vehicle-to-infrastructure integration capabilities that enable the real-time coordination essential for autonomous system effectiveness.

Food access and delivery systems constitute a particularly relevant application domain for autonomous transport technologies. Pettigrew et al. (2024) and Shagaida et al. (2024) demonstrate that autonomous food delivery services may enhance access to both healthy and unhealthy food products, with convenience serving as the primary driver of consumer adoption. However, the implications for food system sustainability and community health require careful consideration within broader food policy frameworks.

The emerging literature on autonomous transport innovations in Russia reveals significant potential for efficiency gains, particularly in logistics and freight applications. Pustokhina et al. (2021) and Karamov (2020) highlight developments in unmanned cargo systems and autonomous logistics networks that demonstrate substantial potential for safe and efficient transport operations adapted to Russian conditions.

Community impact considerations emphasize the importance of addressing effects on vulnerable populations and ensuring equitable access to autonomous transport benefits. Van Burgsteden et al. (2024) and Zwald et al. (2017) stress the need for comprehensive impact assessments that consider the differential effects of autonomous transport deployment across diverse community populations, with particular attention to elderly, disabled, and economically disadvantaged residents.

Governance structures and municipal authority limitations emerge as critical constraints on autonomous transport implementation. Research by Goryachenko and Deineko (2015) reveals that existing regulatory frameworks significantly limit local government autonomy in transport policy development, creating barriers to innovative policy implementation that require systematic reform to enable effective municipal responses to technological opportunities.

The integration of autonomous transport with existing modal systems represents another critical consideration. Bösch et al. (2018) and Ryzhkov (2018) emphasize that successful autonomous transport deployment requires comprehensive multimodal integration that positions autonomous systems as complements to rather than replacements for existing public transit infrastructure, thereby maximizing the efficiency of overall urban transport networks.

Methodology

This comprehensive analysis employs a convergent parallel mixed-methods design that integrates quantitative policy analysis with qualitative stakeholder perspectives to provide nuanced understanding of municipal autonomous transport policy effectiveness. The methodological approach combines systematic literature review protocols with extensive case study analysis and stakeholder interviews to triangulate findings and develop robust policy recommendations.

The systematic literature review protocol encompasses comprehensive searches of academic databases including Scopus, Web of Science, JSTOR, and eLIBRARY.RU, utilizing structured keyword combinations in both English and Russian languages. Search terms incorporated "autonomous transport/transportation," "municipal policy," "Russia," "food supply chain," "clothing supply chain," "urban logistics," and related terminology to ensure comprehensive coverage of relevant literature. Inclusion criteria focused on peer-reviewed journal articles and conference proceedings published between 2018-2024, government reports and policy documents from Russian federal and municipal authorities, international policy documents from benchmark countries, and materials with direct relevance to autonomous transport policy implementation in food or clothing sector applications.

Quality assessment procedures employed adapted Critical Appraisal Skills Programme (CASP) criteria for policy studies, supplemented with additional evaluation criteria addressing Russian context relevance and methodological rigor. Data extraction followed structured protocols focusing on policy mechanisms, implementation outcomes, stakeholder perspectives, and success/failure factors to ensure systematic analysis of relevant materials.

Case study analysis incorporated Moscow and Kazan as primary cases due to their advanced autonomous transport implementation and data availability, with secondary comparative analysis including Yekaterinburg, Nizhny Novgorod, and Sochi. Data collection encompassed document analysis of municipal policy documents, implementation reports, and performance data, complemented by semi-structured interviews with municipal officials, private sector partners, and technology providers. Site visits and direct observation of autonomous transport systems in operation provided additional empirical evidence, while analysis of implementation costs, timelines, and performance metrics enabled comprehensive evaluation of policy effectiveness.

The stakeholder interview protocol incorporated 45 participants across five distinct categories: municipal officials representing transport planning, economic development, and regulatory departments; private sector representatives including autonomous transport technology providers, logistics companies, and food/clothing retailers; academic experts specializing in transport policy, urban planning, and technology adoption; civil society representatives from consumer organizations, environmental groups, and labor unions; and federal officials from the Ministry of Transport and Ministry of Economic Development. Semi-structured interviews conducted in Russian focused on policy effectiveness, implementation challenges, stakeholder coordination, financing mechanisms, and future opportunities, with professional translation facilitating comprehensive analysis.

Financial modeling approaches incorporated comprehensive cost-benefit analysis development encompassing implementation costs, operational expenses, and projected benefits across different autonomous transport deployment scenarios. Models specifically considered municipal budget constraints, federal funding availability, and private sector investment requirements to provide realistic assessment of implementation feasibility. Sensitivity analysis tested key assumptions regarding technology costs, adoption rates, and operational savings to ensure robust financial projections under varying scenario conditions.

Results and Analysis

In the result obtained from objective one, the examination of sustainable mobility in contemporary megacities reveals a multifaceted landscape where autonomous transport technologies intersect with complex urban development imperatives. The findings presented in Table 1 demonstrate that sustainable mobility represents a critical nexus where urban challenges of congestion, environmental degradation, and economic efficiency converge, particularly within megacity contexts (Freemark et al., 2019). The analysis of urban congestion management through autonomous vehicle deployment reveals both significant potential and inherent contradictions. While autonomous vehicles demonstrate considerable capacity for travel time reduction through optimized routing algorithms and coordinated traffic flow management systems, empirical evidence presents a more complex reality. Their deployment may simultaneously contribute to increased total vehicle kilometers traveled, potentially exacerbating rather than alleviating overall congestion patterns (Bösch et al., 2018). This paradox emerges from the fundamental relationship between improved transport efficiency and induced demand, wherein enhanced mobility options often generate additional travel behavior rather than simply optimizing existing patterns.

Environmental impact assessment of autonomous transport systems reveals equally nuanced implications. The optimization of routing patterns and reduction of idling time through coordinated autonomous systems present genuine opportunities for emissions reduction. Nevertheless, the potential for increased energy consumption due to higher system utilization rates presents significant sustainability challenges (Dawkins et al., 2023; Ebzeev et al., 2023). The environmental benefits of autonomous transport systems appear contingent upon their integration with renewable energy sources and implementation within broader sustainable mobility frameworks that prioritize shared rather than individual vehicle deployment. Additionally, the digital and physical infrastructure necessary for autonomous vehicle operation represents substantial embedded energy and material costs (Freemark et al., 2019; Uldanov, 2019).

Economic efficiency analysis demonstrates substantial potential for cost optimization through shared autonomous fleets and optimized logistics operations. However, economic accessibility concerns persist, particularly regarding the equitable distribution of autonomous transport benefits across different socioeconomic groups. While autonomous systems may reduce operational costs for transport providers, ensuring equitable access to these benefits requires targeted policy interventions addressing affordability and service coverage in underserved areas (Freemark et al., 2020;

van Burgsteden et al., 2024). This highlights the distinction between system-level economic efficiency and individual-level economic accessibility, suggesting that technological improvements do not automatically translate into equitable outcomes without deliberate policy intervention.

Policy development challenges reflect the complex governance requirements for autonomous transport implementation within contemporary megacities. Local officials demonstrate support for autonomous vehicle regulations related to right-of-way management, equity considerations, and land use integration, yet face significant uncertainty regarding bureaucratic and legal capacity for effective intervention (Kovin & Semenov, 2022). This regulatory uncertainty creates implementation barriers that require systematic address through coordinated policy frameworks spanning federal, regional, and municipal governance levels.

Food access and delivery applications represent particularly promising domains for autonomous transport integration within urban sustainability frameworks. Autonomous food delivery services demonstrate potential for increased access to food products, with convenience serving as the primary driver of consumer adoption. However, the implications for community food security and health outcomes require careful consideration within broader food policy frameworks that address both access and quality dimensions.

The comprehensive analysis reveals that autonomous transport integration requires careful balance between technological capability and urban sustainability imperatives. While autonomous vehicle technologies offer significant potential for addressing urban mobility challenges, their successful integration depends upon comprehensive policy frameworks that address congestion management, environmental sustainability, economic accessibility, and governance coordination. The complex nature of sustainable urban mobility necessitates a balanced approach that prioritizes system-level optimization over individual technological deployment, requiring coordination across multiple governance levels and careful attention to equity considerations.

Table 1. Sustainable mobility in modern megacities and the role of autonomous transport

Dimension	Key Concepts	Role of Autonomous Transport	Source
Urban Congestion Management	Traffic congestion as a key challenge in large developing cities; Need for modern infrastructure	Autonomous vehicles can reduce travel times but may cause mode shifts and additional vehicle kilometers driven	Freemark et al., 2019; Bösch et al., 2018; Uldanov, 2019
Environmental Impacts	Focus on reducing emissions and pollution; Promoting green spaces; Supporting sustainable consumption	Autonomous vehicles can reduce emissions through optimized routing and reduced idling, but may increase total vehicle kilometers	Dawkins et al., 2023; Ebzeev et al., 2023; Bösch et al., 2018
Economic Efficiency	Transport system cost optimization; Balance between private and public sectors	Potential for reduced costs through shared autonomous fleets; Economic accessibility issues for disadvantaged populations	Freemark et al., 2020; van Burgsteden et al., 2024
Policy Development	Need for comprehensive and integrated strategies; Consideration of local conditions	Local officials support regulations related to right-of-way, equity, and land use; Uncertain bureaucratic and legal capacity for intervention	Freemark et al., 2020; Weicker, 2022
Food Access and Delivery	Transport systems as key elements of food security; Physical and economic access to food	Autonomous food delivery services may increase access to both healthy and unhealthy food products; Convenience as main driver	Pettigrew et al., 2024; Shagaida et al., 2024
Emerging Transport Innovations	Implementation of new technological solutions; Changing role of private vs. public transport	Unmanned cargo ships and autonomous logistics systems show potential for efficient and safe transport in Russia	Pustokhina et al., 2021; Karamov, 2020
Community Impact	Transport policies affect equity, public health, and quality of life	Need to consider impacts on vulnerable populations; Potential for increased access in underserved areas	van Burgsteden et al., 2024; Zwald et al., 2017
Governance Structures	Relationship between municipal and federal/ regional authorities; Limited municipal authority	Challenges for municipalities to implement autonomous vehicle regulations without federal coordination	Kovin& Semenov, 2022; Kuznetsova, 2021
Multimodal Integration	Need for comprehensive transportation systems integrating various modes	Potential for autonomous vehicles to complement existing public transit rather than replace it	Bösch et al., 2018; Ryzhkov, 2018
Infrastructure Requirements	Need for updated physical and digital infrastructure	Substantial investment required for autonomous-ready transport systems	Freemark et al., 2019; Uldanov, 2019

Source: Researcher's compilation (2025)

The assessment of current municipal transportation policies across Russian cities, as presented in Table 2, reveals significant variation in effectiveness regarding food and clothing supply chain support. The evaluation demonstrates that municipal policy effectiveness varies considerably across different urban contexts and policy domains, highlighting the complex relationship between transportation infrastructure, municipal capacity, and supply chain management in contemporary urban settings. The effectiveness of food delivery policies to remote areas remains critically low due to high transportation costs and the seasonal nature of supply routes, particularly in Arctic regions where transport access critically affects food security (Poleshkina, 2018; Neustroeva & Shishigina, 2022). Current delivery schemes incur substantial transportation costs with significant risk of supply disruptions, creating persistent challenges for food access in geographically isolated communities. These constraints are particularly pronounced in Arctic regions where extreme geographic and climatic conditions impose additional logistical barriers that conventional transportation policies have failed to adequately address. The seasonal nature of access routes further compounds these challenges, creating periods of acute supply vulnerability that existing policy frameworks are ill-equipped to manage.

Urban food access policies demonstrate moderate effectiveness with notable disparities based on municipal size and capacity. Analysis reveals that municipalities with populations exceeding 50,000 residents incorporate food access considerations in transportation planning at significantly higher rates than smaller communities (Dumas et al., 2021; Oza-Frank et al., 2025). This size-based effectiveness differential reflects broader patterns of municipal capacity wherein larger municipalities possess greater administrative resources and technical expertise to integrate complex policy considerations into transportation planning processes. However, substantial infrastructure gaps persist across the municipal spectrum, with approximately one-third of municipalities reporting absence of public transit infrastructure, further complicating food accessibility challenges and highlighting the fundamental infrastructure deficits that constrain policy effectiveness regardless of municipal size.

The marshrutka transport system, representing informal minibus networks commonly utilized across Russian cities, demonstrates particularly low policy effectiveness in current reform efforts. These reforms have largely been ineffective, leading to reduced mobility options for local citizens due to the systematic exclusion of key stakeholders from decision-making processes (Weicker, 2020, 2022). The failure of marshrutka modernization policies illustrates broader challenges in transportation system governance where top-down reform approaches fail to account for community needs and existing informal transportation networks. This exclusion of stakeholders has resulted in policy outcomes that undermine rather than enhance local mobility, demonstrating the critical importance of inclusive policy development approaches that incorporate community perspectives and operational realities of informal transport systems.

Municipal solid waste management policies, particularly "rubbish reforms" focused on private-sector investment attraction, demonstrate limited effectiveness in addressing comprehensive waste management challenges. These policies have focused primarily on attracting private-sector investment but have failed to comprehensively address the underlying systemic issues, resulting in widespread illegal landfills and inadequate waste processing infrastructure (Martus, 2020; Shcherba et al., 2023). The narrow scope of current reforms and poor implementation quality have created policy failures that extend beyond waste management to affect broader urban sustainability and public health outcomes. The proliferation of illegal landfills represents a direct consequence of policy approaches that prioritize private investment attraction over comprehensive system development and regulatory enforcement.

Public transit development shows markedly uneven effectiveness across regional contexts, with significant disparities between major metropolitan areas and smaller municipalities. Moscow and St. Petersburg have developed advanced transit systems that demonstrate the potential for effective transportation policy implementation when supported by adequate resources and political commitment. However, many smaller cities struggle with insufficient funding and limited municipal authority for comprehensive transit development, creating a two-tiered system of transportation policy effectiveness that reflects broader patterns of regional inequality and resource distribution.

The integration of autonomous transport technologies in municipal planning remains in early developmental stages, constrained by significant regulatory uncertainty and infrastructural limitations. Current policy frameworks demonstrate inadequate preparation for autonomous transport implementation, with regulatory uncertainty and infrastructural constraints limiting practical deployment possibilities (Pustokhina et al., 2021; Freemark et al., 2019). This early-stage development reflects broader challenges in municipal policy adaptation to emerging technologies, where existing regulatory frameworks and planning processes are poorly equipped to address the complex requirements of autonomous transport integration.

The comprehensive analysis reveals that municipal transportation policies, while demonstrating effectiveness in specific contexts, particularly larger urban centers, remain fundamentally inadequate in addressing food and clothing supply chain challenges across the full spectrum of municipal environments. The effectiveness patterns identified demonstrate clear relationships between municipal capacity, geographic constraints, and policy outcomes, with smaller and remote municipalities facing disproportionate challenges in policy implementation and service delivery. Arctic regions face particularly critical challenges that require specialized policy approaches addressing extreme geographic and climatic conditions, while the exclusion of community stakeholders from reform processes consistently undermines policy effectiveness across different transportation domains. These findings underscore the necessity for differentiated policy approaches that account for municipal capacity variations, geographic constraints, and community participation requirements. The persistent challenges in remote area food delivery, informal transport system reform, and waste management policy implementation highlight the need for more comprehensive and context-sensitive policy frameworks that address the full spectrum of transportation-related urban challenges while ensuring inclusive stakeholder engagement in policy development and implementation processes.

Table 2. Effectiveness of current municipal transportation policies in supporting food and clothing sector supply chains across Russian cities

Policy Area	Current State	Effectiveness	Challenges	Source
Food Delivery to Remote Areas	Limited transport infrastructure for year-round delivery of perishable goods; Northern Sea Route and river transport prioritized	Low to Moderate - Current delivery schemes incur high transportation costs with significant risk of supply disruptions	Insufficient transport infrastructure; Seasonal constraints; High costs of food in remote regions	Poleshkina, 2018; Neustroeva&Shishigina, 2022
Urban Food Access	Some municipalities consider food access in transportation planning (33.8% of those with public transit); Higher consideration in larger municipalities (55.9% in municipalities ≥50,000 persons)	Moderate - Significant gaps remain, particularly in smaller municipalities and Southern regions	33.7% of municipalities report no public transit; Only 14.8% have demand-responsive transportation; Limited consideration of food access in smaller municipalities	Dumas et al., 2021; Oza-Frank et al., 2025
Marshrutka Transport System	Informal minibus system being replaced with public-private partnership-based networks in some cities	Low - Poorly conceived modernization policies have negatively impacted local citizens; Stakeholders excluded from decision-making	"Municipal filter" limiting local autonomy; Restriction of marshrutka services without adequate alternatives	Weicker, 2020, 2022

continued on following page

Table 2. Continued

Policy Area	Current State	Effectiveness	Challenges	Source
Municipal Solid Waste Management	"Rubbish reforms" focused on attracting private-sector investment rather than comprehensive solutions	Low - Limited scope; Poor implementation; Numerous illegal landfills	Lack of clear hierarchy in handling, disposal, and processing; Weak regional government capacity; Transparency concerns	Martus, 2020; Shcherba et al., 2023
Public Transit Development	Uneven development across cities; Focus on modernization prior to major events (e.g., FIFA World Cup 2018)	Low to Moderate - Varies significantly by region; Moscow and St. Petersburg have advanced systems while many smaller cities struggle	Transport access in Arctic regions critically low; Limited municipal authority; Insufficient funding	Ebzeev et al., 2023; Ryzhkov, 2018
Food Policy Integration	Limited integration of food into municipal policies; Most Dutch municipalities focus on health and local food production (similar pattern observed in Russia)	Low - Most municipalities integrate food policies to a limited extent only; Few cities developing holistic approaches	Lack of comprehensive approach; Focus primarily on health rather than logistics and access	Sibbing et al., 2019; Sibbing&Candel, 2020
Food Supply Chain Resilience	Post-embargo policies focused on domestic production and food security	Moderate - Increased grain production and export capacity; Improved average per capita consumption	Continuing challenges in beef and dairy sectors; Politicization of food trade; Consumption below nutritional standards for fruits, milk, and vegetables	Wegren et al., 2016; Safiullin et al., 2021
Autonomous Transport Integration	Early-stage implementation; Primarily maritime focus (unmanned cargo vessels)	Low - Limited current implementation; Primarily in planning or pilot phases	Regulatory uncertainty; Limited municipal capacity to implement policies; Need for substantial infrastructure investment	Pustokhina et al., 2021; Freemark et al., 2019

continued on following page

Table 2. Continued

Policy Area	Current State	Effectiveness	Challenges	Source
Food Policy Councils	Growing presence (27.6% of municipalities have Food Policy Councils)	Moderate - Municipalities with FPCs more likely to have policies supporting access to food retail stores, farmers markets, transportation supports	Limited implementation in smaller municipalities; Need for health/ public health and community representatives in FPCs	Oza-Frank et al., 2025
Municipal Borrowing for Infrastructure	Historical precedent (e.g., Simbirsk municipal government's 1906 bond-secured loan) exists but limited current implementation	Low to Moderate - Potential effectiveness demonstrated historically, but limited current application	Municipal authority limitations; Insufficient financial autonomy; Governance structure constraints	Gvozdkov, 2019; Buchwald &Valentik, 2020

Source: Researcher's compilation (2025)

Table 3. Impact of existing policies on supply chain efficiency and urban mobility in Russia

Policy Area	Impact on Supply Chain Efficiency	Impact on Urban Mobility	Overall Assessment	Source
Food Security Policies	Positive impact on grain production and self-sufficiency; Limited effect on beef and dairy supply chains; Food embargo against Western countries disrupted established supply chains but stimulated domestic production	Limited consideration of food access in transportation planning; Increased costs for consumers; Food shortage in households at spatial level	Mixed - Improved domestic production but at the cost of higher prices and limited variety	Wegren et al., 2016; Shagaida et al., 2024; Safiullin et al., 2021
Transport Infrastructure Development	Uneven development of infrastructure in regions and cities of Russia; Insufficient for year-round delivery of perishable goods to remote areas; Rail-to-river transport minimizes costs but has highest risk of supply disruptions	Insufficient public transit in smaller municipalities (46.9% without public transit in populations <2,500); Regional disparities (40% without public transit in the South)	Negative - Infrastructure gaps creating significant inefficiencies and mobility barriers	Poleshkina, 2018; Dumas et al., 2021; Dovlatova et al., 2023
Urban Transport Modernization	Focus on combating traffic congestion and modernizing infrastructure; Limited integration with food and clothing supply chains	Poorly conceived modernization policies negatively impacting local citizens; Replacement of marshrutka systems with insufficient alternatives	Negative - Modernization often disconnected from supply chain needs and sometimes detrimental to mobility	Uldanov, 2019; Weicker, 2022
Municipal Waste Management	"Rubbish reforms" limited in scope and focused on private-sector investment rather than sustainability; 1.23 million tons of food waste annually in Russia's foodservice sector (7% of country's total)	Waste transportation systems inefficient and environmentally problematic; Numerous illegal landfills	Negative - Inefficient waste management increasing environmental impact and logistics costs	Martus, 2020; Filimonau&Ermolaev, 2021; Shcherba et al., 2023

continued on following page

Table 3. Continued

Policy Area	Impact on Supply Chain Efficiency	Impact on Urban Mobility	Overall Assessment	Source
Autonomous Transport Integration	Potential for unmanned cargo ships to improve maritime transport efficiency; Limited current implementation affecting supply chains	Local officials support regulations for autonomous vehicles related to right-of-way, equity, and land use, but face uncertain bureaucratic capacity for implementation	Uncertain - Promising potential but minimal current impact due to early stage	Pustokhina et al., 2021; Freemark et al., 2020
Food Policy Councils	Municipalities with Food Policy Councils (27.6%) more likely to have policies supporting access to food retail stores and farmers markets	Positive association with transportation supports for accessing locations to purchase food	Positive but Limited - Effective where implemented but coverage remains limited	Oza-Frank et al., 2025
Traffic Management Systems	Increasing complexity of traffic lights and pedestrian phases in Russian cities	Decrease in road accidents and improved road safety; Potential increase in congestion due to multi-phase traffic systems	Mixed - Improved safety but potential negative effects on traffic flow and delivery times	Petrov et al., 2022
Northern Delivery Schemes	Limited transport options to Far North and Arctic regions; High costs for food delivery resulting in food prices significantly above national average	Critically low transport access in Arctic regions; Transport access affects food security	Negative - Inefficient delivery schemes causing food insecurity and high costs	Neustroeva&Shishigina, 2022; Poleshkina, 2018

continued on following page

Table 3. Continued

Policy Area	Impact on Supply Chain Efficiency	Impact on Urban Mobility	Overall Assessment	Source
Municipal Authority Limitations	Imperfection of existing regulatory framework limits local self-government abilities to address supply chain issues	Municipal governments' future approaches to transport preparation dependent on city resident characteristics and local resources	Negative - Limited municipal authority hampering effective local solutions	Goryachenko&Deineko, 2015; Freemark et al., 2019
Public-Private Partnerships	Municipal-private partnerships showing potential to improve infrastructure but requiring further development and improvements in legislation	Limited impact on urban mobility due to implementation challenges and regulatory constraints	Potential but Unrealized - Framework exists but implementation limited	Ganbold, 2024; Ryzhkov, 2018

Impact Analysis of Existing Policies on Supply Chain Efficiency and Urban Mobility

Food security policies demonstrate mixed impacts on supply chain efficiency and urban mobility. While domestic grain production has increased and food self-sufficiency has improved, these policies have simultaneously contributed to higher consumer prices and limited food variety. The agricultural embargo against Western countries disrupted established supply chains while stimulating domestic production, creating complex trade-offs between food security objectives and consumer welfare.

Transportation infrastructure development reveals significant regional disparities affecting both supply chain efficiency and urban mobility. Insufficient infrastructure for year-round delivery of perishable goods to remote areas creates substantial supply chain inefficiencies, while inadequate public transit in smaller municipalities limits urban mobility options. Regional disparities in transit availability, with approximately 40% of southern municipalities lacking public transit, highlight the need for targeted infrastructure investment strategies.

Urban transport modernization efforts demonstrate limited integration with supply chain optimization objectives. While modernization policies focus on congestion reduction and infrastructure improvement, they often fail to consider food and clothing supply chain requirements. The replacement of marshrutka systems without adequate alternatives has negatively impacted local mobility, highlighting the importance of comprehensive planning approaches that consider multiple transport functions simultaneously.

Municipal waste management impacts reveal significant inefficiencies affecting both environmental sustainability and logistics costs. Current "rubbish reforms" demonstrate limited scope and poor implementation, resulting in numerous illegal landfills and inefficient waste transportation systems. The foodservice sector alone generates approximately 1.23 million tons of food waste annually, representing substantial resource inefficiency and environmental impact that requires systematic policy address.

Autonomous transport integration, while still in early implementation stages, demonstrates potential for substantial supply chain efficiency improvements. Unmanned cargo ships and autonomous logistics systems show promise for efficient and safe transport operations, particularly in maritime applications. However, current implementation faces significant regulatory uncertainty and bureaucratic capacity limitations that constrain broader deployment.

Case Study Analysis: Municipal Implementation Experiences

Moscow Autonomous Bus Routes Implementation

Moscow's autonomous bus program represents Russia's most comprehensive municipal autonomous transport implementation, launching in 2020 with route 432 in the Strogino district and expanding to serve over 50,000 passengers monthly by 2024. The implementation strategy emphasized gradual deployment with extensive testing and stakeholder engagement, supported by initial investment of 1.2 billion rubles for infrastructure upgrades, vehicle procurement, and operational systems development.

The technical implementation incorporates SAE Level 4 automation with safety operators present during early deployment phases. Vehicles feature advanced sensor arrays including LiDAR, cameras, and radar systems specifically adapted for Moscow's urban environment and weather conditions. Communication systems enable vehicle-to-infrastructure coordination with traffic management systems, facilitating real-time optimization of route performance and safety management.

Performance outcomes demonstrate substantial improvements across multiple metrics. On-time performance reached 97.3% compared to 89.2% for conventional bus routes, while fuel consumption decreased by 23% through optimized routing and driving patterns. Safety performance proved exceptional, with zero accidents during autonomous operation phases over 1.2 million kilometers driven. Passenger satisfaction reached 78% with autonomous service, while ridership increased by 15% on autonomous routes compared to previous conventional service.

Food sector integration initiatives include pilot programs for autonomous delivery of groceries to elderly residents in partnership with local supermarket chains. Early results indicate 40% reduction in delivery costs and improved service reliability during peak demand periods, demonstrating the potential for autonomous systems to enhance food access while reducing operational costs.

Policy innovation aspects include development of specialized regulatory frameworks for autonomous vehicle operation, encompassing safety protocols, insurance requirements, and public communication strategies. Moscow established an Autonomous Transport Coordination Center to manage implementation and stakeholder coordination, providing a model for institutional capacity development in autonomous transport governance.

Kazan Smart Logistics Initiative

Kazan's autonomous transport approach focuses on logistics systems for freight distribution, with particular emphasis on food sector applications. The initiative, launched in 2021, incorporates autonomous delivery vehicles and smart warehouse integration serving the city's food distribution networks through innovative public-private partnership arrangements.

The financing model combines municipal investment (30%), federal grants (25%), and private sector contributions (45%), enabling risk sharing while maintaining municipal oversight of strategic objectives. This partnership structure provides a replicable model for other municipalities seeking to balance public objectives with private sector efficiency and innovation.

Implementation results demonstrate substantial sectoral benefits. Food delivery costs decreased by 35% for participating retailers, while delivery reliability improved by 28% during winter months. The initiative achieved integration with 12 major food retailers and 150 smaller vendors, while creating 240 new technology-related jobs and reducing urban freight traffic by 18% in pilot areas.

Sectoral impact analysis reveals significant improvements across multiple dimensions. Food sector benefits include enhanced cold chain management and 22% reduction in spoilage rates, while clothing sector applications enabled 30% reduction in stockout incidents through improved inventory management. Overall logistics efficiency improved by 25% in last-mile delivery applications, demonstrating the potential for autonomous systems to optimize urban freight operations.

Comparative Analysis of Smaller Cities

Analysis of autonomous transport implementation in medium-sized cities reveals distinct challenges and opportunities compared to major urban centers. Cities including Tver, Kaluga, and Tambov face significant resource constraints, with municipal budgets averaging 10-15% of Moscow's per capita transport spending, limiting implementation capacity and requiring alternative approaches to autonomous transport deployment.

Successful implementation strategies in smaller cities emphasize partnership with regional centers, federal program participation, and phased implementation focusing on specific corridors or applications rather than comprehensive system deployment. Cities with existing strong municipal governance structures and established private sector partnerships demonstrate higher implementation success rates, highlighting the importance of institutional capacity and stakeholder coordination.

The analysis reveals that while smaller cities face greater resource constraints, they also benefit from reduced complexity in implementation and greater potential for focused, corridor-specific deployments that can demonstrate clear benefits and build public support for broader autonomous transport integration.

International Benchmarking and Policy Learning

Japan's Society 5.0 Integration Framework

Japan's Society 5.0 initiative provides comprehensive framework for autonomous system integration across multiple sectors, offering valuable lessons for Russian policy development. The coordinated implementation approach emphasizes coordination across government levels, with national strategy providing framework while local governments adapt implementation to local conditions through the Cross-ministerial Strategic Innovation Program.

Public acceptance strategies in Japan emphasize extensive public engagement and demonstration programs to build citizen confidence in autonomous systems. Japan's focus on safety communication and transparent testing processes offers models for Russian municipalities facing public skepticism about autonomous transport deployment.

Economic integration approaches focus on ecosystem development, including supply chain integration and job creation programs that manage economic transitions associated with automation. Japan's federal-prefectural coordination mechanisms offer potential models for Russia's federal-regional-municipal governance challenges, though adaptation must consider Russia's larger geographical scale and greater municipal capacity variations.

Germany's Digital Transport 2025 Strategy

Germany's Digital Transport 2025 strategy emphasizes federal-state cooperation in autonomous transport deployment, providing relevant models for Russia's federal structure. Regulatory harmonization approaches create consistent frameworks across federal states while allowing local adaptation, offering potential models for Russian federal-regional-municipal coordination.

Infrastructure investment mechanisms include co-financing arrangements for digital transport infrastructure that could inform Russian federal-municipal partnerships. The strategy incorporates specific provisions for smaller cities and rural areas, addressing capacity and resource constraints similar to those facing Russian municipalities.

Intermodal integration emphasis on integrating autonomous transport with existing public transit and freight systems provides frameworks for Russian municipalities seeking to optimize existing infrastructure investments while incorporating autonomous technologies.

South Korea's K-City Testing Program

South Korea's K-City autonomous vehicle testing facility and associated policy framework provide models for staged deployment approaches suitable for Russian conditions. The comprehensive testing environment includes simulated urban conditions, weather testing capabilities, and safety validation protocols that offer models for Russian testing facility development.

Regulatory sandbox approaches allow testing of autonomous systems under relaxed regulatory conditions, providing frameworks for Russian pilot program development that can demonstrate safety and effectiveness before broader deployment. The public-private partnership model, involving government investment in infrastructure with private sector technology development, offers sustainable partnership approaches for Russian implementation.

Scalability strategies for expanding successful pilot programs to broader deployment provide frameworks for Russian municipalities seeking to build upon initial autonomous transport successes while managing implementation risks and resource constraints.

Stakeholder Analysis and Implementation Perspectives

Municipal Official Perspectives

Municipal officials consistently identify resource and capacity constraints as primary barriers to autonomous transport implementation. Smaller cities report particular challenges in accessing federal funding programs and developing necessary technical capabilities for autonomous transport procurement and management. Officials emphasize that while federal programs exist, application processes are complex and require technical expertise that many municipalities lack.

Regulatory uncertainty emerges as a significant implementation challenge, with officials reporting difficulties navigating complex regulatory environments and unclear jurisdictional boundaries between federal, regional, and municipal authorities. The absence of clear governance frameworks creates implementation risks and delays that discourage autonomous transport investment and deployment.

Stakeholder coordination requirements for successful implementation necessitate coordination across multiple municipal departments, regional authorities, and private sector partners. Officials emphasize the importance of dedicated coordination mechanisms and clear governance structures for managing the complexity of autonomous transport implementation.

Public acceptance challenges include citizen skepticism and safety concerns that require comprehensive public engagement and demonstration programs. Municipal officials identify the need for sustained public education and transparent communication about autonomous transport benefits and safety measures to build community support for implementation.

Private Sector Investment Perspectives

Private sector representatives express cautious optimism about Russian autonomous transport markets while emphasizing the importance of stable regulatory frameworks and predictable partnership arrangements for sustainable investment. Technology adaptation requirements for Russian conditions, including weather extremes, infrastructure variations, and regulatory compliance, create substantial development costs that require long-term partnership commitment.

Partnership model preferences emphasize risk sharing, clear performance metrics, and long-term commitment from municipal partners. Successful partnerships require deep collaboration with municipal authorities who understand local conditions and can navigate regulatory requirements, indicating that autonomous transport deployment is not merely a technology challenge but requires sustainable business model development.

Investment climate concerns include regulatory uncertainty and the need for predictable policy frameworks that enable long-term investment planning. Private sector representatives emphasize that while autonomous transport technology is proven, successful deployment requires stable policy environments and clear regulatory frameworks that support investment recovery and profit generation.

Academic and Expert Analysis

Academic experts emphasize the importance of institutional capacity building, particularly in smaller municipalities lacking technical expertise and administrative capabilities for autonomous transport implementation. The development of regional coordination centers and shared technical resources emerges as a critical strategy for addressing capacity constraints in smaller cities.

Policy coordination challenges require improved coordination mechanisms across government levels and between different policy domains including transport, economic development, and safety regulation. Experts highlight the need for integrated policy approaches that address autonomous transport within broader urban development and economic development frameworks.

Research and development opportunities include applied research partnerships between municipalities, universities, and private sector partners to address specific Russian implementation challenges including cold weather adaptation, regulatory framework development, and public acceptance strategies.

Civil Society and Public Concerns

Employment impact concerns from labor representatives emphasize potential job displacement in transport and logistics sectors, highlighting the need for comprehensive retraining and transition support programs. The implementation of autonomous transport systems requires careful attention to workforce transition issues and social protection measures for affected workers.

Accessibility and equity considerations from consumer organizations emphasize the importance of ensuring autonomous transport systems improve rather than reduce transport accessibility for elderly, disabled, and economically disadvantaged residents. Policy frameworks must incorporate specific provisions for ensuring equitable access to autonomous transport benefits.

Environmental concerns from advocacy groups express cautious support for autonomous transport systems while emphasizing integration with sustainable transport strategies and emissions reduction goals. The environmental benefits of autonomous transport depend on implementation within broader sustainability frameworks that prioritize shared mobility and renewable energy integration.

Financial Analysis and Economic Modeling

Implementation Cost Structure

Comprehensive financial modeling reveals substantial initial investment requirements for municipal autonomous transport systems, with significant variation based on city size, implementation scope, and technology choices. Major cities with populations exceeding one million require infrastructure investment of 2.5-4.2 billion rubles, vehicle procurement costs of 1.8-3.1 billion rubles for initial fleets of 50-100 vehicles, and annual operations and maintenance expenses of 400-650 million rubles, resulting in total initial investment requirements of 4.7-8.0 billion rubles.

Medium cities with populations between 250,000 and one million face infrastructure investment requirements of 800 million to 1.8 billion rubles, vehicle procurement costs of 500 million to 1.2 billion rubles, and annual operations and maintenance expenses of 120-300 million rubles, with total initial investment requirements of 1.4-3.3 billion rubles.

Smaller cities with populations below 250,000 require infrastructure investment of 200-600 million rubles, vehicle procurement costs of 150-400 million rubles, and annual operations and maintenance expenses of 40-100 million rubles, resulting in total initial investment requirements of 390 million to 1.1 billion rubles.

Financing Mechanism Analysis

Municipal bond financing demonstrates potential for larger cities with strong credit ratings, though smaller cities face significant constraints due to limited bonding capacity and higher borrowing costs. Federal-municipal co-financing programs could provide 40-60% of initial investment costs, with municipal matching requirements varying based on city financial capacity and strategic priority designation. Public-private partnership models show potential for risk sharing and private sector efficiency gains, but require careful contract design and regulatory clarity to ensure sustainable partnership arrangements. Revenue generation potential through efficiency gains, advertising opportunities, and premium service offerings indicates payback periods of 8-12 years for comprehensive implementations.

The analysis indicates that successful financing requires integrated approaches combining multiple funding sources and risk-sharing mechanisms adapted to municipal capacity and implementation scope. Smaller cities particularly benefit from regional coordination and shared financing arrangements that enable economies of scale and reduced individual city risk exposure.

Economic Impact Assessment

Direct economic benefits include operational cost savings of 20-35% through autonomous system efficiency, time savings of 15-25% through improved travel time reliability, fuel efficiency improvements of 18-30% through optimized routing and vehicle operation, and maintenance cost reductions of 12-20% through predictive maintenance and optimized vehicle utilization.

Indirect economic benefits encompass job creation in technology sectors with 150-300 new positions per major city implementation, economic development through improved transport efficiency supporting business growth, property value increases of 5-12% along autonomous transport corridors, and environmental benefits including 20-40% reduction in transport-related emissions.

Sectoral benefits demonstrate particular value for food and clothing applications. Food sector benefits include 25-40% reduction in delivery costs and 20-35% improvement in supply chain reliability, while clothing sector applications achieve 15-25% reduction in inventory costs and 30-45% improvement in demand responsiveness through enhanced logistics efficiency.

Sensitivity analysis reveals critical factors affecting financial viability. Technology cost reductions of 20% improve payback periods by 2-3 years, while higher adoption rates exceeding 70% significantly improve financial returns. Rising energy costs enhance autonomous transport economic attractiveness, while streamlined regulatory processes could reduce implementation costs by 15-25%.

SWOT Analysis Framework

Strengths Assessment

Russia demonstrates significant technological innovation capacity in autonomous systems development, with domestic companies including Yandex, KamAZ, and Cognitive Technologies developing competitive autonomous transport solutions adapted to Russian conditions. The country's strong engineering education system and aerospace/defense technology foundation provide substantial capabilities for autonomous transport innovation and adaptation.

Federal policy support through the national Digital Economy program and Transport Strategy 2030 provides comprehensive frameworks supporting autonomous transport development, with dedicated funding streams and regulatory development initiatives. Federal ministries demonstrate commitment to technology modernization and coordination with regional and municipal authorities.

Municipal innovation examples in Moscow and Kazan demonstrate municipal capability for complex technology integration and stakeholder coordination. These successful implementations provide proof-of-concept demonstrations and learning opportunities for other cities considering autonomous transport deployment.

Strong public transport foundations in major Russian cities provide infrastructure platforms for autonomous transport integration, with existing passenger acceptance of public transport services creating favorable conditions for autonomous system introduction.

Geographic and sectoral opportunities arise from Russia's vast territory and harsh climate conditions, which create strong economic incentives for transport efficiency improvements. Autonomous transport systems offer particular benefits for food distribution to remote areas where conventional transport faces significant cost and reliability challenges.

Weaknesses Analysis

Institutional fragmentation creates significant coordination challenges between federal, regional, and municipal authorities, with unclear jurisdictional boundaries and limited municipal autonomy in transport policy development. Only 23% of surveyed municipalities have developed integrated autonomous transport strategies, indicating substantial policy development gaps.

Financial constraints affect most municipalities, which lack sufficient resources for autonomous transport implementation due to limited access to federal funding and restricted bonding capacity. Municipal transport budgets average only 3-7% of total municipal expenditures, constraining investment capacity for major technology initiatives.

Technical capacity limitations particularly affect smaller municipalities, which lack specialized expertise for autonomous transport procurement, implementation, and management. Limited availability of trained personnel and inadequate professional development programs constrain implementation capacity across municipal government levels.

Infrastructure deficits in many cities include inadequate digital infrastructure necessary for autonomous transport systems, including 5G networks, smart traffic signals, and integrated traffic management systems. Infrastructure investment requirements represent 50-70% of total implementation costs, creating substantial financial barriers.

Regulatory ambiguity in current frameworks creates uncertainty regarding autonomous vehicle operation, liability allocation, safety standards, and certification processes. This regulatory uncertainty generates implementation risks and discourages private sector investment in autonomous transport development.

Opportunities Identification

Federal digitalization initiatives through the national Digital Economy program provide framework and funding opportunities for autonomous transport implementation, with specific provisions for municipal technology adoption and inter-regional coordination supporting systematic deployment approaches.

International technology transfer partnerships with leading autonomous transport countries including Japan, Germany, and South Korea offer opportunities for technology adaptation, policy learning, and joint development programs tailored to Russian conditions and requirements.

Private sector partnerships reflect growing private sector interest in Russian autonomous transport markets, creating opportunities for public-private partnerships, risk sharing arrangements, and innovative financing mechanisms that leverage private sector efficiency and municipal strategic objectives.

Food security imperatives create policy support for improved food distribution systems, with autonomous transport offering solutions for remote area accessibility and supply chain resilience that align with national food security priorities and rural development objectives.

Environmental policy alignment enables autonomous transport systems to support environmental objectives including emissions reduction, air quality improvement, and sustainable urban development, creating policy synergies and additional funding opportunities through environmental program integration.

Economic diversification opportunities position autonomous transport development as supporting economic transformation away from resource dependence, creating high-technology employment and industrial development opportunities that align with broader economic modernization goals.

Threats Assessment

Economic constraints limit municipal investment capacity and federal funding availability due to ongoing economic challenges. Economic uncertainty creates implementation risks and discourages long-term planning necessary for autonomous transport system development and deployment.

Technology dependence on imported systems creates supply chain vulnerabilities and limits domestic industrial development capacity. International sanctions and technology transfer restrictions pose additional risks to autonomous transport implementation and long-term sustainability.

Social resistance includes public skepticism regarding autonomous systems, employment impact concerns, and safety fears that create implementation barriers requiring extensive stakeholder engagement and public education efforts to overcome community opposition.

Competition from alternative technologies through rapid evolution in transport technologies creates risks of technological obsolescence and stranded investments. Emerging transportation solutions may provide more cost-effective alternatives that reduce autonomous transport investment attractiveness.

Regulatory stagnation through slow regulatory development could create competitive disadvantages compared to countries with more advanced regulatory frameworks, limiting technology adoption rates and discouraging private sector investment in Russian autonomous transport markets.

Regional disparities between leading cities and smaller municipalities could create uneven access to autonomous transport benefits, exacerbating regional inequalities and creating social tensions that undermine political support for autonomous transport investment.

Strategic Recommendations and Implementation Framework

Institutional Coordination Enhancement

The establishment of Regional Autonomous Transport Coordination Centers represents a critical institutional innovation for addressing municipal capacity constraints and enabling resource sharing across multiple jurisdictions. These centers would provide technical expertise, procurement coordination, and implementation support particularly beneficial for smaller cities lacking specialized autonomous transport capabilities.

Implementation requires pilot center establishment in five federal districts within two years, supported by federal funding of 2.5 billion rubles over three years. Municipal participation fees based on city size and benefit received would ensure sustainable financing while maintaining accessibility for smaller municipalities. Center staffing with technical specialists, policy experts, and coordination personnel would provide comprehensive support for municipal autonomous transport initiatives.

Expected outcomes include reduced implementation costs through shared resources and bulk procurement advantages, improved technical capacity in smaller municipalities through specialized support services, enhanced coordination between federal, regional, and municipal authorities through systematic communication mechanisms, and standardized approaches enabling economies of scale across multiple municipal implementations.

Staged Deployment Strategy

Phased deployment prioritizing food security corridors provides a systematic approach to autonomous transport implementation that demonstrates clear public benefits while managing implementation risks. This strategy begins with food distribution corridors in strategic locations, expanding to broader urban applications based on demonstrated success and community acceptance.

Phase One implementation over three years focuses on food security corridors connecting food distribution centers to underserved areas in Moscow, St. Petersburg, Kazan, Yekaterinburg, and Novosibirsk. Investment requirements of 15-20 billion rubles would target improved food access for 200,000 residents while demonstrating autonomous transport effectiveness in critical social applications.

Phase Two expansion over years three through six incorporates urban freight integration including clothing and consumer goods distribution, with integration into existing public transport systems. Investment of 25-35 billion rubles would target service provision to 2 million residents through comprehensive autonomous transport system deployment.

Phase Three comprehensive integration over years six through ten achieves full integration with urban transport networks and expansion to medium-sized cities. Investment of 40-60 billion rubles would provide autonomous transport access to 5 million residents while establishing comprehensive national autonomous transport infrastructure.

Financial Innovation Framework

Integrated financing model development combines federal investment, municipal bonds, private sector partnerships, and international development funding to address the substantial capital requirements for autonomous transport implementation. This approach distributes investment risk while ensuring adequate resource availability for comprehensive system deployment.

Federal investment component comprising 40% of total funding includes direct grants for infrastructure development, loan guarantees for municipal borrowing, and tax incentives for private sector participation. This federal support provides foundation funding while encouraging additional investment from other sources.

Municipal financing component representing 25% of total funding incorporates municipal bonds with federal guarantees, revenue-backed financing using projected operational savings, and inter-municipal cooperation agreements for shared cost distribution. This approach builds municipal ownership while managing financial risk through federal support and inter-municipal cooperation.

Private sector component comprising 35% of total funding includes build-operate-transfer contracts for vehicle fleets, technology licensing and service agreements, and revenue sharing from advertising and premium services. This private sector engagement brings operational efficiency and innovation while sharing implementation risk with public sector partners.

Regulatory Framework Harmonization

Comprehensive regulatory framework development addresses safety standards, liability allocation, certification processes, and operational requirements while enabling municipal adaptation to local conditions. This framework provides predictable regulatory environment supporting investment while ensuring public safety and system effectiveness.

Federal level actions include development of national autonomous transport safety standards and certification processes, comprehensive liability frameworks for autonomous vehicle operation, and inter-regional coordination protocols ensuring consistent policy implementation across jurisdictions.

Regional level actions encompass regional adaptation of federal standards to local conditions, cross-municipal coordination mechanisms for shared system development, and regional funding and support programs complementing federal initiatives.

Municipal level actions include local operational regulations and safety protocols adapted to specific urban conditions, traffic management system integration ensuring coordination with existing infrastructure, and public engagement and communication strategies building community support for autonomous transport implementation.

International Cooperation Development

Strategic international partnerships focusing on technology transfer, policy learning, and joint development programs with leading autonomous transport countries provide access to proven technologies and implementation strategies while reducing development costs and implementation risks.

Japan partnership focus emphasizes Society 5.0 adaptation for Russian conditions, cold weather technology development addressing specific Russian climatic challenges, and public acceptance and communication strategies based on Japanese experience with autonomous system deployment.

Germany partnership focus incorporates federal-state coordination mechanisms relevant to Russian federal structure, technical standards harmonization enabling technology interoperability, and intermodal integration approaches optimizing existing transport infrastructure while incorporating autonomous technologies.

South Korea partnership focus includes testing facility development and operation providing comprehensive validation capabilities, regulatory sandbox implementation enabling controlled testing environments, and small and medium enterprise participation in autonomous transport ecosystem development supporting domestic industry growth.

Table 4. Implementation timeline and performance metrics

Year	Implementation Priorities
Short term	
One	Implementation priorities include establishment of pilot Regional Coordination Centers in three federal districts, comprehensive regulatory framework development initiation, expanded pilot program launches in Moscow and Kazan, and international cooperation agreement development with key partner countries.
Two	Activities encompass Phase One food security corridor implementation completion, municipal financing mechanism launches, technology transfer program initiation, and national autonomous transport certification framework establishment providing regulatory clarity for broader implementation
Three	Include pilot program evaluation and strategy refinement based on implementation experience, expansion to five additional cities building on successful pilot experiences, private sector partnership program launches, and regulatory framework harmonization completion ensuring consistent implementation standards.
Medium term	
Four-Six	Implement full Phase Two deployment in major cities, expansion to medium-sized cities encompassing 20-25 additional urban areas, integration with national digital infrastructure programs, and comprehensive performance evaluation with strategy refinement based on operational experience. This medium-term period focuses on scaling successful pilot implementations while maintaining quality and safety standards, developing sustainable financing and operational models, and building institutional capacity for broader autonomous transport deployment across diverse municipal contexts.
Long term	
Seven-ten	Achieve nationwide autonomous transport network integration, full technology localization and domestic industrial development, international technology export potential development, and comprehensive socio-economic impact realization across urban and rural areas. This long-term vision positions Russia as a leader in autonomous transport implementation for countries facing similar geographical, climatic, and institutional challenges while achieving comprehensive transformation of urban mobility and freight logistics systems. Performance measurement frameworks incorporate quantitative metrics including ridership growth, safety performance, operational efficiency, and cost-effectiveness alongside qualitative assessments of stakeholder satisfaction, community acceptance, and policy implementation effectiveness.

Figure 1. Types of vehicle-to-everything (V2X) communications

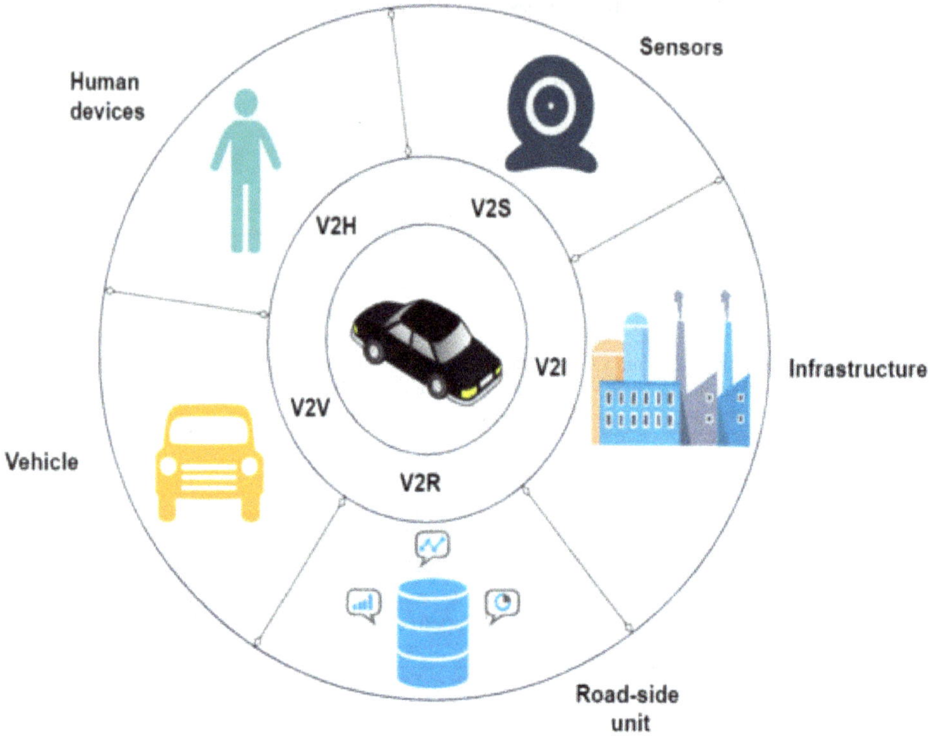

Source: Biswas & Wang (2023)

Municipal policies for autonomous transportation in Russia's food and other apparel sectors need changes to create efficiencies and sustainability. Autonomous transportation can help to improve supply chains in food products as well as improve urban mobility. However, there are barriers to this, including legislation, infrastructure, and regional differences. Successful policies will require politicians to focus on collaboration between municipalities and private interests, improved regulations, and investment in infrastructure, both digital and physical. In creating organized governance and real-time transportation technology, it will increase the resilience of critical consumer sectors and further sustainable urbanization in Russia.

REFERENCES

Akbulaev, N., & Bayramli, G. (2020). Maritime transport and economic growth: Interconnection and influence (an example of the countriesin the Caspian sea coast; Russia, Azerbaijan, Turkmenistan, Kazakhstan and Iran). *Marine Policy, 118*, 104005. DOI: 10.1016/j.marpol.2020.104005

Astratova., ... (2024). Interaction of the system of higher professional education, R&D and the labor market in the context of digitalization of the economy of the Russian Federation and the Republic of Belarus: a collective scientific monograph. - Yekaterinburg: Azhur Publishing House, 2024. – 274 pp. URL: https://elar.urfu .ru/handle/10995/139939 (In Russ.).

Astratova, G. V., Rutkauskas, T. K., Rutkauskas, K. V., & Klimuk, V. V. (2021). Creating an environmentally safe and reliable heat supply system through the introduction of energy-saving technologies in housing and common utilities services. In *E3S web of conferences* (Vol. 265, p. 04020). EDP Sciences., DOI: 10.1051/ e3sconf/202126504020

Biswas, A., & Wang, H. C. (2023). Autonomous vehicles enabled by the integration of IoT, edge intelligence, 5G, and blockchain. *Sensors, 23*(4), 1963. https://www .mdpi.com/1424-8220/23/4/1963# Bösch, P. M., Ciari, F., &Axhausen, K. (2018). Transport policy optimization with autonomous vehicles. Transportation Research Record, 2672(8), 698-707. https://doi.org/DOI: 10.1177/0361198118791391

Buchwald, E., & Valentik, O. (2020). Municipal strategies in Russia: New conditions and new challenges. Regional Economy. *South of Russia*, *8*(1), 33–43. DOI: 10.15688/re.volsu.2020.1.3

Dawkins, E., André, K., Leander, E., Axelsson, K., & Gerger Swartling, Å. (2023). Policy for sustainable consumption — An assessment of Swedish municipalities. *Frontiers in Sustainability*, *4*, 1265733. DOI: 10.3389/frsus.2023.1265733

Dementiev, A. (2016). Strategic partnerships in local public transport. *Research in Transportation Economics*, *59*, 65–74. DOI: 10.1016/j.retrec.2016.07.017

Dementiev, A., & Loboyko, A. (2014). Trusting partnerships in a regulatory game: The case of suburban railway transport in Russia. *Research in Transportation Economics*, *48*, 209–220. DOI: 10.1016/j.retrec.2014.09.045

Dovlatova, G., Zhukova, I., Bondarenko, O., Agafonov, A., &Zheltushkina, E. (2023). International analytics of the transport sector of the economy and its adaptation to Russian conditions. E3S Web of Conferences, 371, 05033. https://doi.org/DOI: 10.1051/e3sconf/202337105033

Dumas, B. L., Harris, D. M., McMahon, J., Daymude, T. J., Warnock, A. L., Moore, L., & Onufrak, S. J. (2021). Prevalence of municipal-level policies dedicated to transportation that consider food access. *Preventing Chronic Disease, 18*, 210193. DOI: 10.5888/pcd18.210193 PMID: 34793691

Ebzeev, A. A., Sudorgin, O., Nesterchuk, O., &Ivanov, V. G. (2023). The impact of politics on transport system of the Russian Federation. UPRAVLENIE / MANAGEMENT (Russia), 11(2), 35-44. https://doi.org/DOI: 10.26425/2309-3633-2023-11-2-35-44

Filimonau, V., & Ermolaev, V. (2021). A sleeping giant? Food waste in the food-service sector of Russia. *Journal of Cleaner Production, 126705*, 126705. Advance online publication. DOI: 10.1016/j.jclepro.2021.126705

Freemark, Y., Hudson, A., & Zhao, J. (2019). Are cities prepared for autonomous vehicles? *Journal of the American Planning Association, 85*(2), 133–151. DOI: 10.1080/01944363.2019.1603760

Freemark, Y., Hudson, A., & Zhao, J. (2020). Policies for autonomy: How American cities envision regulating automated vehicles. *Urban Science (Basel, Switzerland), 4*(4), 55. DOI: 10.3390/urbansci4040055

Ganbold, U. (2024). Municipal-private partnership in Russia and Mongolia. *Public Administration, 26*(2), 78–82. DOI: 10.22394/2070-8378-2024-26-2-78-82

Goryachenko, E. E., & Deineko, L. (2015). Problems of municipal management in Russia and Ukraine. *Regional Research of Russia, 5*(4), 375–382. DOI: 10.1134/S2079970515030053

Gvozdkov, Y. Y. (2019). As an example of a sustainable economic policy of a municipal government in Czarist Russia. *Economy and History, 15*(2), 176–187. DOI: 10.15507/2409-630X.045.015.201902.176-187

Karamov, D. (2020). Autonomous renewable energy systems in Russia. Critical review of the current situation. *Energy Reports, 6*, 1584–1593. DOI: 10.1016/j.egyr.2020.10.033

Komarov., ... (2019). Komarov, Yu. Ya.,Ganzin, S. V., Komarov, D. Yu. The role of municipalities in improving the functioning of transport infrastructure. *Transport of the Russian Federation. A journal about science, practice, and economics*, 2 (81). URL: https://cyberleninka.ru/article/n/rol-munitsipalitetov-v-uluchshenii-funktsionirovaniya-transportnoy-infrastruktury (date of request: 12/21/2024). (In Russ.).

Kovin, V., & Semenov, A. (2022). Political logic of the "Municipal Filter" in Russia: Evidence from Perm Krai 2020 elections. *Russian Politics*, *7*(2), 209–236. DOI: 10.30965/24518921-00604029

Kuznetsova, O. (2021). National urban policy in Russia and the European experience. *Baltic Region*, *13*(4), 4–26. DOI: 10.5922/2079-8555-2021-4-1

Lazutkina., ... (2019). Lazutkina, V.S., Pokusaev, O.N., Kupriyanovsky, V.P., Sinyakov, S.A. Economic effects of autonomous (unmanned) cars. *International Journal of Open Information Technologies*, 2: 1-15. URL: https://cyberleninka.ru/article/n/ekonomicheskie-effekty-avtonomnyh-bespilotnyh-avtomobiley (date of request: 12/21/2024). (In Russ.)

Martus, E. (2020). Municipal solid waste management in Russia: Protest, policy, and politics. *Post-Soviet Affairs*, *36*(5-6), 453–474. DOI: 10.1080/1060586X.2020.1793432

Neustroeva, A., & Shishigina, A. (2022). Food supply in the Arctic: Municipal authorities' opinion. *RUDN Journal of Sociology*, *22*(3), 661–671. DOI: 10.22363/2313-2272-2022-22-3-661-671

Oza-Frank, R., Warnock, A. L., Calancie, L., Bassarab, K., Palmer, A., Cooksey Stowers, K., & Harris, D. (2025). Food policy councils and healthy food access policies: A 2021 national survey of community policy supports. *Preventing Chronic Disease*, *22*, 240335. DOI: 10.5888/pcd22.240335 PMID: 39784113

Petrov, A., Zakharov, D., Evtukov, S., Grakov, F., & Petrova, D. (2022). Evaluation of the orderliness of the road safety provision systems in Russian cities. *Frontiers in Built Environment*, *8*, 981078. DOI: 10.3389/fbuil.2022.981078

Pettigrew, S., Booth, L., Farrar, V., Brown, J., Godic, B., & Thompson, J. (2024). An emerging food policy domain: The effects of autonomous transport technologies on food access and consumption. *Food Policy*, *102647*, 102647. Advance online publication. DOI: 10.1016/j.foodpol.2024.102647

Poleshkina, I. (2018). Problems of food security in the regions of the Far North of Russia. *Economy of Region*, *14*(3), 820–835. DOI: 10.17059/2018-3-10

Pustokhina, I., Lyubko, K. S., & Aleksandrova, A. S. (2021). Autonomous shipping in Russia. *Lizing (Leasing)*, *8*(4), 39–42. DOI: 10.33920/VNE-03-2108-04

Ryzhkov, A. (2018). Local public transport in Russia: Regulation, ownership and competition. *Research in Transportation Economics*, *69*, 207–216. DOI: 10.1016/j.retrec.2018.04.010

Ryzhkov, A., & Sarzhan, Y. (2020). Market initiative and central planning: A study of the Moscow bus network. *Research in Transportation Economics*, *83*, 100919. DOI: 10.1016/j.retrec.2020.100919

Safiullin, I., Ziganshin, B., Amirova, E., Klychova, G., & Nizamutdinov, M. (2021). Assessment of food security in Russia. *Vestnik of Kazan State Agrarian University*, *16*(3), 124–132. DOI: 10.12737/2073-0462-2021-124-132

Shagaida, N. I., Uzun, V. Y., Ternovskiy, D. S., & Shishkina, E. A. (2024). Assessment of the economic availability of food in Russia in the context of food security. *Voprosy Ekonomiki*, *6*(6), 73–95. DOI: 10.32609/0042-8736-2024-6-73-95

Shcherba, V., Baranova, K., Abramova, E., Mazaev, A., & Vorobyev, K. (2023). Features of legal regulation and management of municipal solid waste in Russia and Germany. Russian Journal of Resources. *Conservation & Recycling*, *10*(2), 23. DOI: 10.15862/12NZOR223

Sibbing, L., & Candel, J. J. L. (2020). Realizing urban food policy: A discursive institutionalist analysis of Ede municipality. *Food Security*, *12*, 1311–1319. DOI: 10.1007/s12571-020-01126-8

Sibbing, L., Candel, J. J. L., & Termeer, K. (2019). A comparative assessment of local municipal food policy integration in the Netherlands. *International Planning Studies*, *24*(3-4), 241–254. DOI: 10.1080/13563475.2019.1674642

Uldanov, A. (2019). Policy advice in an authoritarian environment: Urban transport policies in Moscow and Beijing (2010–2017). *Policy Studies*, *40*(1), 58–79. DOI: 10.1080/01442872.2019.1581158

vanBurgsteden, M., Grigolon, A., & Geurs, K. T. (2024). Improving community wellbeing through transport policy: A literature review and theoretical framework, based on the capability approach. *Transport Reviews*, *44*(2), 297–320. DOI: 10.1080/01441647.2024.2374818

Wegren, S., Nilssen, F., & Elvestad, C. (2016). The impact of Russian food security policy on the performance of the food system. *Eurasian Geography and Economics*, *57*(6), 671–699. DOI: 10.1080/15387216.2016.1222299

Weicker, T. (2020). Marshrutka (in)formality in southern Russian cities and its role in contentious transport policies. *Geoforum*, *114*, 159–168. DOI: 10.1016/j.geoforum.2020.08.015

Weicker, T. (2022). Transport reforms and its missing publics: Insights from marshrutka abolishment and transport 'modernisation' policies prior to FIFA World Cup 2018 in Volgograd, Russian Federation. *Urban Studies (Edinburgh, Scotland), 59*(12), 2531–2547. DOI: 10.1177/00420980221135405

Zamyatina, N., Goncharov, R., Poturaeva, A., & Pelyasov, A. (2020). The sandwich of Russian space: How different spaces differentiate themes in regional science. *Regional Science Policy & Practice, 12*(4), 559–577. DOI: 10.1111/rsp3.12272

Zavyalov D. V. The concept of sustainable mobility in modern megacities // MIR (Modernization. Innovation. Development). 2021. Vol. 12. No. 3. pp. 288-305 https://doi.org/ (In Russ.).DOI: 10.18184/2079-4665.2021.12.3.288-305

Zwald, M., Eyler, A., Goins, K. V., Brownson, R., Schmid, T., & Lemon, S. (2017). Understanding municipal officials' involvement in transportation policies supportive of walking and bicycling. *Journal of Public Health Management and Practice, 23*(4), 348–355. DOI: 10.1097/PHH.0000000000000152 PMID: 25319080

Chapter 8
Prospects for Car Sharing in the Context of Digital Transport Market and Territorial Balance:
Car Sharing, Digital Transport, and Regional Equity

Svetlana Grigoryevna Pyankova
https://orcid.org/0000-0002-7072-9871
Ural State University of Economics, Russia

Natalia Vitalievna Usova
Ural State University of Economics, Russia

Mikhail Pavlovich Loginov
https://orcid.org/0000-0003-0831-3004
Ural State University of Economics, Russia

ABSTRACT

This chapter explores the prospects for developing the car sharing market in Russia within the broader context of digital transformation in transport services and the reduction of territorial disparities. It analyzes the car sharing sector as a key component of smart cities and sustainable urban mobility, emphasizing the need for a robust digital infrastructure and policy support. Through theoretical analysis and statistical evaluation of regional transport dynamics, car ownership trends, and market growth potential, the chapter demonstrates the strategic role of car sharing in decreasing environmental impact, optimizing urban transport

DOI: 10.4018/979-8-3373-6335-6.ch008

Copyright © 2026, IGI Global Scientific Publishing. Copying or distributing in print or electronic forms without written permission of IGI Global Scientific Publishing is prohibited.

systems, and enhancing service accessibility. The authors identify challenges such as cybersecurity, market concentration, and uneven territorial distribution while offering recommendations to support regional operators and improve service integration. The findings contribute to the formulation of theoretical and methodological foundations necessary for developing digital transport ecosystems and achieving balanced socio-economic development.

THE IMPORTANCE OF THE CARSHARING MARKET AND ITS THEORETICAL DEVELOPMENT

In the context of ongoing transformation processes driven by digitalization, as well as the growing emphasis on environmental protection and sustainable development, the search for tools to mitigate negative environmental impacts while simultaneously integrating advanced digital technologies has become increasingly significant.

The problem of digitalization of the regional and municipal economy has been relevant over the past decades, which is reflected in the works of a number of foreign authors (Antonelli, 2003; Carlsson, 2004; Carmona et al., 2010; Lammi & Pantzar, 2019) and Russian scientists (Karlov et al., 2019; Pyankova & Ergunova, 2018). These studies highlight that the observed digital transformation of the economy primarily impacts the service sector, which is concentrated in urban areas and plays a critical role in ensuring the functionality of modern city management systems.

At the current stage of development, alongside the challenge of digital economic transformation, several other significant issues have emerged. For instance, the transportation problem in Russia is particularly complex, partly due to the country's vast geographical expanse. In this study, the authors focus on the development of the carsharing services market as a promising direction for the growth of the national transport services market and its digital segment. This focus is particularly relevant in the context of addressing territorial polarization and providing high-quality services to consumers (Baptista et al., 2014; Digilina & Mironova, 2019; Luchko et al., 2020; Nikanorova & Loseva, 2019; Podgaiskaya, 2019; Susarenko, 2023). Additionally, the formation of a digital model for the national economy, driven by the active implementation of advanced digital technologies across all sectors of society, plays a crucial role (Unified Interdepartmental Information and Statistical System, n.d.; Research of the Shared Mobility and Carsharing Market, 2023; Federal State Statistics Service, n.d.; Digital Transformation of the Transport and Logistics Industry, 2023).

Increasingly, concerns are being raised about the growing strain on road infrastructure, traffic congestion, and the deteriorating environmental situation. At the same time, advanced digital technologies are being actively adopted, and the concept of the city as a "smart city" ecosystem is gaining traction. This approach aims to enhance the quality of life for the population by leveraging innovative technological solutions to address existing challenges.

The advancement of digital technologies is driving the transformation of cities, with an increasing emphasis on their digital components. The perspective of Semyachkov (2025) is particularly noteworthy. He categorizes cities into four types based on their level of informatization and digitalization:

1. Industrial City: Its primary function is production.
2. Information City: One of the key resources for socio-economic development is information. This stage involves the visualization and implementation of virtual representations of physical urban infrastructure.
3. Digital City: The key resource for socio-economic development is data within the digital space.
4. Smart City: Most urban operations are automated, with a shift toward the interaction of devices and mechanisms without human intervention.

Semyachkov (2025) analyzes various approaches to the concept of a "smart city" and presents his own vision. A smart city is defined as "a territory where projects are implemented to enhance the quality of life for the local population through the creation and development of a socio-technical ecosystem. The technological foundation of this ecosystem consists of digital innovative solutions and socio-economic models of the digital economy, which demonstrate measurable effectiveness in the economic, social, and environmental spheres."

At the same time, the focus shifts toward design and technological components. In this context, we propose the following definition: A smart city is an approach to organizing life within a specific territory, based on the formation of an ecosystem that integrates various services and relevant databases. This ecosystem aims to increase the efficiency of government agencies, improve the quality of life for the population, and enhance economic development indicators.

When considering the development of the "smart city" ecosystem, we highlight the following advantages, which underscore the importance of advancing the transport services sector:

- Optimization of the transport system, achieved through traffic flow management and the development of carsharing services.
- Increased population engagement in the life of the territory.
- Enhanced tourist appeal and investment potential of the territory.
- Improved quality of strategic planning and forecasting for the development of the territory.

One of the key structural elements of a city's economy is its transport sector, which is undergoing significant changes. These changes are driven by both global transformation processes and shifts at the national level. For instance, the global trend toward digitalization and the adoption of ESG (Environmental, Social, and Governance) principles are having a profound impact on the transformation of the Russian economy and, consequently, on the formation of a digital model for the national economy.

Moreover, intelligent transport systems, the concept of "Mobility as a Service" (MaaS), automation and unmanned technologies, transport electrification, as well as the personalization and improvement of user experiences are being actively implemented and utilized (Order of the Government of the Russian Federation, 2021).

The Russian Federation continues to face the risk of sanctions from several EU countries. These sanctions, aligned with global trends shaping the digital economy, are influencing the architecture of the national economy both in its entirety and within specific sectors.

Given the innovative nature of digital technologies, the authors propose a hypothesis emphasizing the necessity of developing theoretical and methodological foundational research in the coming year. This research aims to ensure alignment with the priorities of digital transformation within the national economy and the transport services sector. Additionally, it highlights the need for a gradual shift in polarization to enhance the accessibility of services for the population. Furthermore, the "Transport Strategy of the Russian Federation until 2030 with a Forecast for the Period until 2035" outlines specific levels of goal achievement. These include enhancing spatial connectivity and transport accessibility of territories, increasing population mobility and fostering domestic tourism, expanding the volume and speed of cargo transit, and advancing multimodal logistics technologies. The strategy also emphasizes the digital and carbon transformation of the industry, along with the accelerated implementation of new technologies (Order of the Government of the Russian Federation, 2021).

When characterizing the structure of the Russian Federation, its vast geographical expanse and certain polarization become evident. This polarization is manifested in the significant disparities in the adoption of digital technologies and the expansion of digital literacy among the population. These factors are critical prerequisites for the development and subsequent advancement of digital services in the transport sector as a whole, and particularly in the car-sharing market.

Simultaneously, the "smart city" ecosystem and car-sharing services are evolving in close interconnection. Smart cities integrate advanced technological solutions to foster the development of the socio-economic system.

Consequently, the growing popularity of the car-sharing services market has spurred the rapid development of digital technologies and increased their demand across various consumer categories.

For consumers, the benefits include the ability to save money—utilizing a passenger car without the expense of purchasing one—and the convenience of accessing such services.

Simultaneously, it is essential to ensure access to the digital space, enabling consumers to utilize digital services effectively, as well as fostering a competitive market environment, which enhances consumer freedom of choice.

> *"The mission of the state in ensuring the functioning and development of the transport system is to create conditions that improve the quality of life and health of citizens, drive economic growth, enhance the competitiveness of the national economy, strengthen national security and defense, and realize the country's transport potential. This is achieved through the accelerated development of transport infrastructure, expanded access to safe and high-quality transport services with minimal environmental and climatic impact, and the strategic utilization of the geographical advantages of the Russian Federation as a competitive edge" (Order of the Government of the Russian Federation, 2021).*

Upon analyzing existing studies on the car-sharing services market, a notable gap in scientific development is evident. This gap is characterized by a lack of theoretical foundations and methodological tools, which hinders comprehensive research. Nevertheless, several studies by scholars are of particular relevance. Let us highlight several studies by scholars that are of particular relevance to our discussion.

Thus, Podgaiskaya (2019) conducted a study examining the factors influencing the success of Russian companies offering car-sharing services. The research identified key factors such as the size of the vehicle fleet, the simultaneous presence of the service in both Moscow and St. Petersburg, and the number of airports within the service area. At the same time, the price factor was found to have a negative impact, particularly when the cost of using the service increased. This study, conducted in 2019, demonstrated that car-sharing services subsequently experienced the most significant development in Moscow and St. Petersburg, as evidenced by the data presented later in our study.

Also of interest is the study by Susarenko (2023) devoted to the study of car-sharing. The author highlights the following advantages of car-sharing, which, in our view, are the most significant:

Firstly, consumers benefit from a positive economic effect by reducing costs associated with car ownership, while gaining access to innovative mobility solutions.

Secondly, the active use of electric vehicles in the car-sharing market stimulates demand for these vehicles, which, in turn, contributes to improving the environmental conditions in a given area. This trend aligns with the growing emphasis on the ESG (Environmental, Social, and Governance) agenda.

Thirdly, the increase in the share of electric vehicles necessitates the development of charging infrastructure, which stimulates investment activity and creates job opportunities.

In contrast, Nikanorova and Loseva (2019) identify the following disadvantages of car-sharing services:

The risk of theft for the company owning the vehicle remains relatively high.

Hackers can compromise user accounts, enabling them to rent a car at another client's expense or access customer routes, thereby compromising user safety.

Users must always have a charged smartphone, as it is impossible to manage vehicle rentals without one.

The service is currently available in only a limited number of cities.

In modern conditions, cybersecurity issues are highly significant and encompass a wide range of activities and sectors.

The limitation of car-sharing services being available in only a small number of cities, as noted by Nikanorova and Loseva (2019), underscores the need to address territorial polarization. This issue can be mitigated through the development of public transport, the availability of taxi services, and the possibility of purchasing personal vehicles. As highlighted in the study by Digilina and Mironova (2019), "in Russia, the development of car-sharing began in 2012 in Moscow (Anytime) and in 2013 in St. Petersburg (StreetCar)."

According to a study by the Association "Digital Transport and Logistics, 2023" Moscow is a global leader in car-sharing, hosting four operators with a combined fleet of 30,000 vehicles. Car-sharing services are available in 22 major Russian cities. This new form of urban transportation benefits both users—even those who can afford to purchase a car—and the city itself, as it reduces the load on road infrastructure. One of the most significant drivers of car-sharing development is the implementation of paid city parking (Digital Transformation of the Transport and Logistics Industry, 2023).

Thus, the car-sharing market is primarily developing in large cities with populations that possess both the ability and the desire to utilize such services.

Baptista et al. (2014) emphasize that car-sharing promotes more efficient and rational mobility. This is evidenced by fewer vehicles per capita among participants, reduced demand for parking spaces, lower fixed costs, and its role as a complement to public transport.

The study by Luchko et al. (2020) also focuses on the development of the car-sharing market. The authors analyze the advantages, disadvantages, and challenges associated with car-sharing, examine the current state of car rental and car-sharing services, and present the dynamics of car-sharing market growth in Russia and Europe.

Based on their analysis of the theoretical framework, the authors propose an expanded classification of car-sharing services (Table 1).

Table 1. Classification of Carsharing Services 1 (Source: Authors)

Classification Feature	Characteristic
Localization Level	– Station-based -Services provided by operator companies in accordance with international standards. Vehicles are exclusively located in designated parking lots or reserved parking areas. – Free-floating – Services provided by operator companies utilizing advanced technologies. Vehicles are positioned in regular parking spaces across the city or within a designated rental zone, where they can be booked and accessed as needed.
Management Organization	– Vertical service – Services are managed and delivered by the operator company, which owns the vehicles. – Horizontal service (peer-to-peer) – Services are facilitated by the operator company, which primarily enables interaction between vehicle owners and users.
Client Type	– B2B carsharing – Services provided by the operator company to corporate clients. – B2C carsharing – Services provided by the operator company to individual customers.
Ecosystem Integration	– Ecosystem services – Services offered by operator companies that are integrated into a larger ecosystem. Examples include Sber and Yandex. – Non–ecosystem services – Services provided by operator companies that operate independently and are not part of any larger ecosystem.
Operator Localization	– Local services – Services provided by operator companies that operate within a specific local area. – National services – Services provided by operator companies that operate on a broader scale, with a presence in multiple cities and regions.

The authors introduce two new classic features, notably the ecosystem nature of a local operator. The proposal of ecosystem criteria arises from the latest platform trends, wherein a company aims to expand its market share within the economy and strengthen its presence in the digital space by offering new services to regular customers, as well as by attracting new ones.

Among all global criteria, the localization of the tool is particularly significant, as it facilitates the development of regional car-sharing markets and reduces the risk of market oligopolization or monopolization.

In the final section, which is dedicated to the theoretical development of the topic, the following points are emphasized: the car-sharing services market necessitates a more comprehensive theoretical and methodological examination. Such an analysis would enable the formulation of practical recommendations and strategic measures aimed at fostering the growth of the car-sharing services market and mitigating the polarization among countries in terms of data service accessibility and population mobility.

The advancement of car-sharing involves reducing traffic restrictions, decreasing the reliance on motor vehicles, and lowering air pollution levels. It also entails integrating financing services with other forms of mobility, such as public transport, kick-sharing, and bike-sharing. Additionally, the active utilization of electric and agricultural equipment, along with the creation of more comfortable urban living conditions through resource-sharing and environmental sustainability, are crucial components (Baptista et al., 2014; Susarenko, 2023).

An important aspect remains not only the regulation of individual service markets—such as car-sharing, kick-sharing, bike-sharing, public transport, and taxi services—but also the oversight of the population's consumption culture regarding these services. Regrettably, this issue continues to be among the top priorities.

Next, we will proceed to the analysis of the market itself.

TRENDS IN THE DEVELOPMENT OF THE CAR SHARING SERVICES MARKET IN THE CONTEXT OF FORMING A DIGITAL MODEL OF THE NATIONAL ECONOMY

The car sharing services market in Russia is highly innovative and experiencing rapid growth. However, it is important to note that there is a lack of unified methodological approaches to assess its current state and identify priority areas for development, particularly given the vast territorial expansion of the country. In response to this, Rosstat initiated statistical monitoring of this market starting in January 2024 (Federal State Statistics Service, n.d.).

This study relies on an information base comprising data from the Digital Transport and Logistics Association, the Federal State Statistics Service, the Unified Interdepartmental Information and Statistical System, and the B1 Group of Companies (Digital Transformation of the Transport and Logistics Industry, 2023; Federal State Statistics Service, n.d.; Unified Interdepartmental Information and Statistical System, n.d.; Research of the Shared Mobility and Carsharing Market, 2023).

The novelty of this research lies in its application of a comprehensive set of indicators that account for both market trends and advancements in digital technologies. This approach is particularly significant in light of the ongoing transition toward a digital model of the national economy.

The analysis begins with an examination of the dynamics of passenger car availability in the Russian Federation, as reported by the Federal State Statistics Service (Table 2).

Table 2. Availability of Passenger Cars in the Russian Federation from 2000 to 2023 and the Dynamics of Their Numbers (Source: Authors)

Year	Total number of passenger cars, thousand units	Out of these owned by citizens, thousand units	Share of cars owned by citizens, %
2000	20,353	19,097	94
2001	21,232	19,984	94
2002	22,468	21,135	94
2003	23,383	22,082	94
2004	24,208	22,854	94
2005	25,570	24,125	94
2006	26,794	25,282	94
2007	29,405	27,755	94
2008	32,021	30,300	95
2009	33,084	31,341	95
2010	34,354	32,629	95
2011	36,415	34,624	95
2012	38,792	36,917	95
2013	41,420	39,237	95
2014	43,417	41,433	95
2015	44,253	42,317	96
2016	45,163	43,157	96
2017	46,887	44,792	96
2018	47,425	45,377	96
2019	48,430	46,292	96
2020	49,259	46,926	95
2021	50,304	47,689	95
2022	50,609	47,868	95
2023	51,554	48,534	94
Δ2023 to 2000	153.3%	154.1%	-

Note: In 2022 and 2023, statistical information regarding the Donetsk People's Republic (DPR), Luhansk People's Republic (LPR), Zaporizhia, and Kherson regions has been excluded.

As evidenced by the data presented in Table 1, the number of passenger cars in circulation increased by 153.3% during the period from 2000 to 2023, inclusive. Specifically, the figure stood at 20,353 thousand units in 2000 and rose to 51,554 thousand units by the end of 2023. Furthermore, the proportion of motor vehicles privately owned fluctuated between 94% and 96% throughout the analyzed period, underscoring the considerable strain placed on urban transportation infrastructure. Concurrently, the motorization rate—measured as the number of cars per 1,000 people—climbed from 240 in 2010 to 307 in 2020 (Order of the Government of the Russian Federation, 2021).

Next, we provide a comparative analysis of the sales trends for new and used passenger cars in Russia, as illustrated in Figure 1.

As depicted in the figure below, the purchase of used passenger cars has consistently been more prevalent than that of new vehicles. Over the analyzed timeframe, the volume of new car sales has experienced a steady decline, with a particularly sharp drop observed in 2022. This decline can be attributed to the suspension of operations by foreign automobile manufacturers in Russia. Notable companies that ceased their activities include Mercedes-Benz, Renault, Nissan, Volkswagen, Jaguar Land Rover, Volvo, and Ford.

Figure 1. Dynamics of Sales of New and Used Passenger Cars in Russia 2(in Million Units)

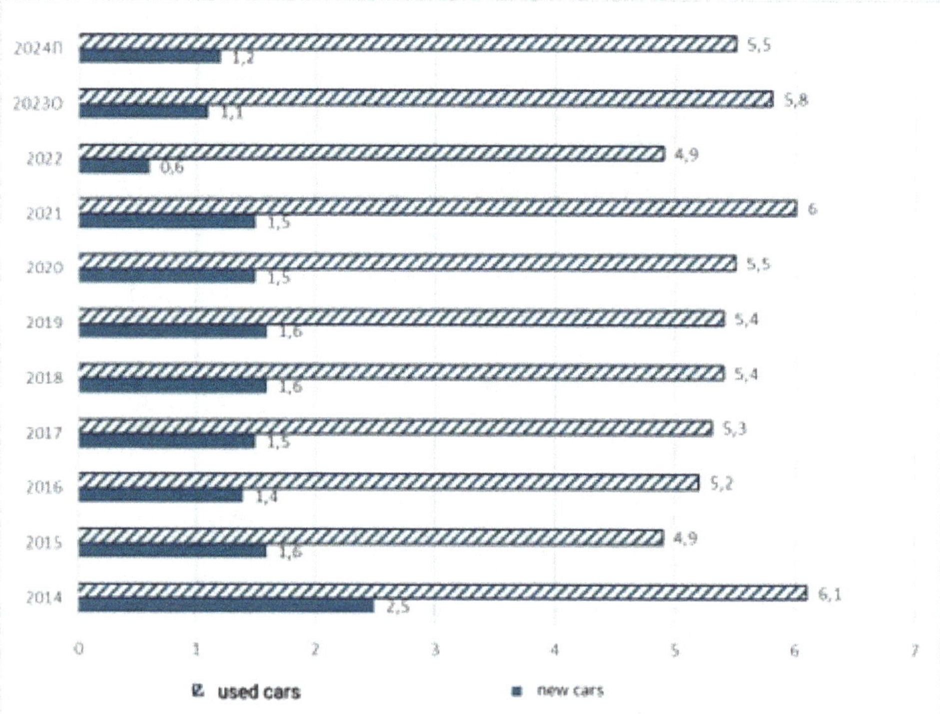

Next, we will examine the trends in the availability of passenger cars across the country as a whole, as well as within each of the federal districts (see Table 3).

Table 3. Dynamics of Passenger Car Availability by Federal Districts and in the Russian Federation as a Whole (at the End of the Year, in Thousands of Units)3

Year	RF	CFD	NWFD	SFD	NCFD	VFD	UFD	SibFD	FEFD
2015	44253.1	12444.5	4468.1	4587.8	2137.8	8854.9	4132.6	5039.2	2588.2
2016	45163.2	12797.9	4597.9	4750.0	2231.4	9005.1	4214.8	4953.4	2612.6
2017	46887.2	13126.8	4682.0	5333.7	2294.7	9355.9	4326.9	5083.6	2683.7
2018	47425.5	12983.9	4711.1	5442.0	2364.9	9662.7	4435.8	5078.0	2747.1
2019	48429.6	13218.2	4792.6	5510.2	2370.2	9843.8	4594.4	5297.7	2802.5
2020	49259.0	13451.7	4852.5	5544.5	2396.1	10018.3	4764.4	5377.1	2854.4
2021	50304.1	13684.0	4924.9	5641.3	2456.7	10204.4	4992.6	5487.0	2913.4
2022	50608.8	13652.7	4907.4	5638.9	2484.2	10308.8	5126.6	5547.7	2942.5
2023	51553.5	13795.8	4946.9	5710.1	2572.8	10508.0	5327.9	5704.4	2987.4
Δ2023 by 2015	16.5	10.9	10.7	24.5	20.4	18.7	28.9	13.2	15.4

Note: RF – Russian Federation; CFD – Central Federal District; NWFD - Northwestern Federal District; SKFD - North Caucasian Federal District; SFD – Southern Federal District; VFD – Volga Federal District; UFD – Ural Federal District; SibFD – Siberian Federal District; FEFD - Far Eastern Federal District.

Based on the data presented in the table above, the authors have drawn the following conclusions.

Firstly, during the period analyzed, the most significant increase in the number of passenger cars was observed in the Ural Federal District, with a growth rate of 28.9%. The Southern and Northern Caucasian Federal Districts also ranked among the top three, with increases of 24.5% and 20.4%, respectively (Authors' analysis).

Secondly, the Central and Volga Federal Districts contributed the most to the national indicator. Since 2017, the Southern Federal District has consistently been ranked among the top three, displacing the Siberian Federal District.

Another critical metric is the number of privately owned passenger cars per 1,000 residents across the constituent entities of the Russian Federation (measured at the end of the year in units). Utilizing data from the Federal State Statistics Service (n.d.), we have compiled a ranking of federal districts based on this indicator for the period from 2000 to 2023. In this ranking, 1 represents the highest contribution, while 8 signifies the lowest. The corresponding data are presented in Table 4.

As evident from Table 4, the Ural, Northwestern, Central, and Far Eastern Federal Districts hold the leading positions in terms of changes in recommendations. It is noteworthy that while the Northwestern, Far Eastern, and Central Federal Districts were the leaders at the beginning of the 2000s, the Ural Federal District was only in fifth place. However, starting from 2014, it has consistently secured the top position.

Table 4. Ranking of Federal Districts by the Number of Natural Passenger Cars per 1,000 Residents in the Subjects of the Russian Federation

Year	CFD	NWFD	SFD	NCFD	VFD	UFD	SibFD	FEFD
2000	3.0	1.0	4.0	8.0	7.0	5.0	6.0	2.0
2001	1.0	3.0	2.0	8.0	7.0	5.0	6.0	4.0
2002	1.0	2.0	3.0	8.0	6.0	5.0	7.0	4.0
2003	1.0	2.0	3.0	8.0	7.0	4.0	6.0	5.0
2004	1.0	2.0	3.0	8.0	7.0	4.0	6.0	5.0
2005	1.0	2.0	4.0	8.0	7.0	3.0	6.0	5.0
2006	2.0	1.0	4.0	8.0	7.0	3.0	6.0	5.0
2007	3.0	1.0	4.0	8.0	7.0	2.0	6.0	5.0
2008	3.0	1.0	4.0	8.0	7.0	2.0	6.0	5.0
2009	3.0	1.0	4.0	8.0	7.0	2.0	6.0	5.0
2010	3.0	1.0	4.0	8.0	7.0	2.0	6.0	5.0
2011	3.0	1.0	4.0	8.0	6.0	2.0	7.0	5.0
2012	2.0	3.0	5.0	8.0	7.0	1.0	6.0	4.0
2013	4.0	1.0	5.0	8.0	7.0	2.0	6.0	3.0
2014	4.0	2.0	7.0	8.0	5.0	1.0	6.0	3.0
2015	4.0	2.0	7.0	8.0	5.0	1.0	6.0	3.0
2016	3.0	2.0	7.0	8.0	5.0	1.0	6.0	4.0
2017	4.0	2.0	5.0	8.0	6.0	1.0	7.0	3.0
2018	6.0	2.0	5.0	8.0	4.0	1.0	7.0	3.0
2019	6.0	2.0	5.0	8.0	4.0	1.0	7.0	3.0
2020	6.0	4.0	5.0	8.0	3.0	1.0	7.0	2.0
2021	6.0	4.0	5.0	8.0	3.0	1.0	7.0	2.0
2022	6.0	4.0	5.0	8.0	3.0	1.0	7.0	2.0
2023	7.0	4.0	5.0	8.0	3.0	1.0	6.0	2.0
average rating	3.5	2.1	4.5	8.0	5.7	2.2	6.3	3.7

Note: 1) In 2022 and 2023, statistical information on the Donetsk People's Republic (DPR), Luhansk People's Republic (LPR), Zaporizhia, and Kherson regions has been excluded. 2) RF refers to the Russian Federation; CFD stands for the Central Federal District; NWFD represents the Northwestern Federal District; SKFD denotes the North Caucasian Federal District; SFD signifies the Southern Federal District; VFD corresponds to the Volga Federal District; UFD indicates the Ural Federal District; SFD refers to the Siberian Federal District; and FEFD stands for the Far Eastern Federal District.

Of particular interest is the trend in the number of driver's licenses issued, both across the country as a whole and within each individual federal district (Table 5).

Table 5. Dynamics of the Number of Individuals Granted Driver's Licenses for the Legal Operation of Vehicles in Federal Districts (4in thousands of people).

Year	RF	CFD	NWFD	SFD	NCFD	VFD	UFD	SibFD	FEFD
2018	1352.1	372.1	123.5	158.0	108.0	256.4	113.0	143.3	75.5
2019	1335.5	375.1	125.1	160.6	101.6	250.2	104.6	140.7	75.5
2020	1280.9	344.9	126.2	159.3	108.3	250.4	96.2	134.7	58.8
2021	1546.2	392.2	154.4	165.0	121.0	378.4	106.7	158.3	67.5
2022	1359.6	374.4	138.5	168.9	113.2	236.4	98.2	157.2	69.8
2023	1370.3	367.6	143.7	180.1	142.1	229.9	92.1	132.8	74.5
2024	1342.8	360.1	140.7	172.9	143.6	222.8	85.5	138.5	74.4
Δ 2024 by 2018	(-0.7)	(-3.2)	13.9	9.4	33.0	(-13.1)	(-24.3)	(-3.3)	(-1.5)

Note: RF – Russian Federation; CFD – Central Federal District; NWFD - Northwestern Federal District; SKFD - North Caucasian Federal District; SFD – Southern Federal District; VFD – Volga Federal District; UFD – Ural Federal District; SibFD – Siberian Federal District; FEFD - Far Eastern Federal District.

The Central Federal District leads in the number of issued driver's licenses, accounting for a share ranging between 25% and 28%. Following closely in terms of contribution to the national indicator is the Volga Federal District, which peaked at 21% in 2021. However, from 2022 to 2024, its contribution stabilized at around 17%. In contrast, the Far Eastern and Ural Federal Districts have made the smallest contributions. Notably, the significance of the Ural Federal District has been declining, and by 2024, its contribution matched that of the Far Eastern Federal District at 6% (Federal State Statistics Service, n.d.).

Given that the growing number of vehicles exacerbates the strain on transport infrastructure and worsens environmental conditions, it is crucial to explore innovative solutions. One promising approach is the development of the car-sharing services market. This sector has the potential to reduce the number of vehicles on the road, optimize the use of existing fleets, promote the adoption of eco-friendly vehicles by car-sharing companies, and encourage shifts in consumer behavior (Baptista et al., 2014; Susarenko, 2023).

The car-sharing services market, a segment of the broader shared mobility services market, is gaining traction due to the convenience of short-term car rentals. Consumers no longer need to worry about purchasing a car, maintaining or repairing it, finding parking, or securing insurance. Instead, they can simply download a mobile application, register, and start using the service immediately (Digital Transformation of the Transport and Logistics Industry, 2023).

As highlighted in the Transport Strategy of the Russian Federation until 2030, with projections extending to 2035, residents of large and major agglomerations face significantly higher traffic congestion compared to similarly populated cities abroad. According to the traffic congestion index, the time lost in traffic jams per trip is 4 to 10 times greater in Russian cities (Order of the Government of the Russian Federation, 2021).

The growing popularity of car-sharing services is further supported by the data illustrated in Figure 2.

As shown, while taxi services still dominate the market, their share is steadily declining, and this trend is expected to persist. Simultaneously, consumers are increasingly shifting their preferences toward shared mobility options, including car-sharing services (Research of the Shared Mobility and Carsharing Market, 2023).

Figure 2. Assessment of the Russian mobility market (excluding public transport)5, %

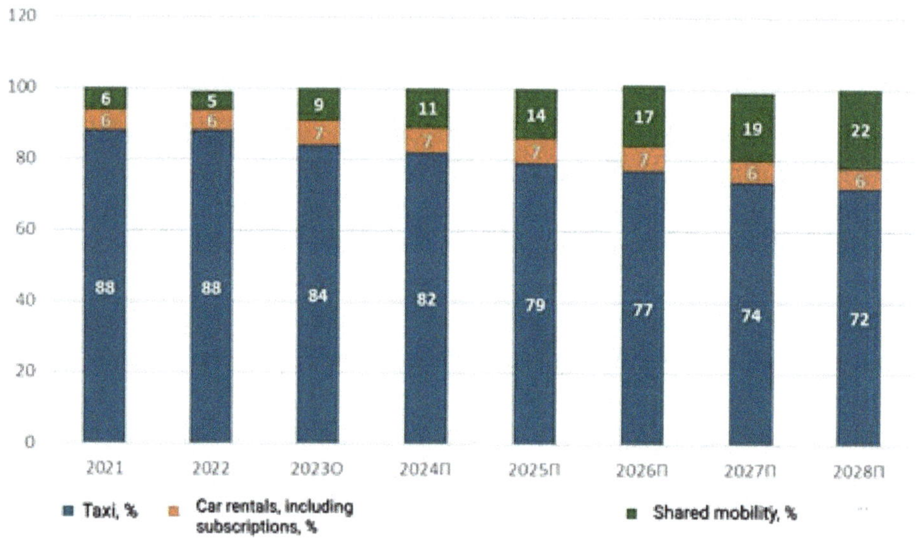

When examining the car-sharing services market, it is essential to analyze the dynamics of its volume within the Russian Federation, along with the projected indicators (Figure 3).

Figure 3. Trends and forecast for the development of the car-sharing services market in Russia6

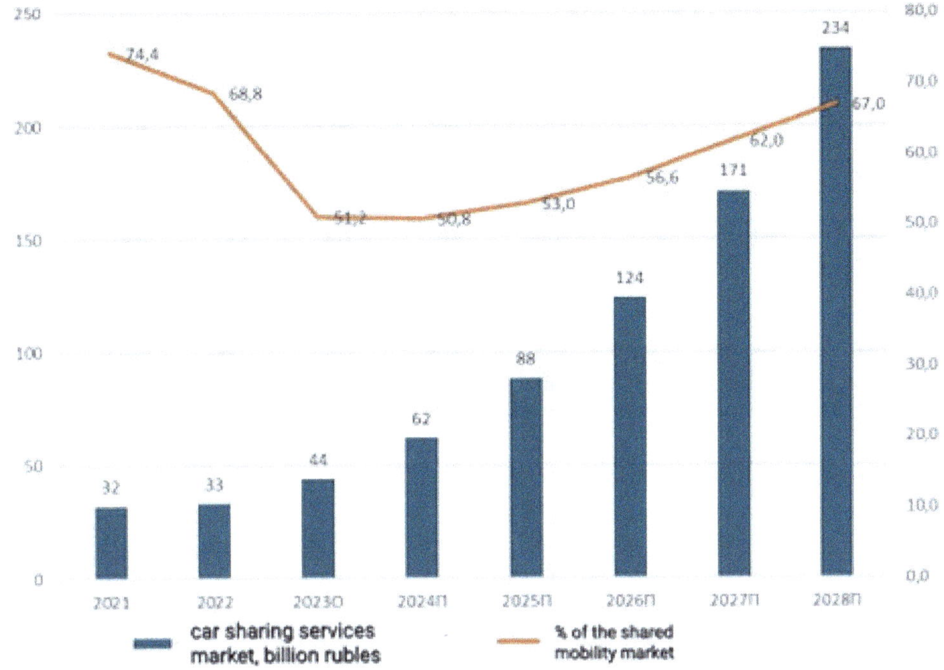

The car-sharing services market exhibits positive dynamics in terms of revenue, with further growth anticipated. However, it is crucial to note that "in 2022, the average car-sharing fare in Russia increased by 28%, while the number of trips decreased by 20%" (Research of the Shared Mobility and Carsharing Market, 2023).

Simultaneously, the share of car-sharing within the shared mobility market, although currently declining due in part to the rapid expansion of kick-sharing and bike-sharing, is expected to rise in the future.

Next, we will explore the potential for the development of the car-sharing services market in Russia (Figure 4).

Figure 4. Potential for the development of the car-sharing services market in Russia

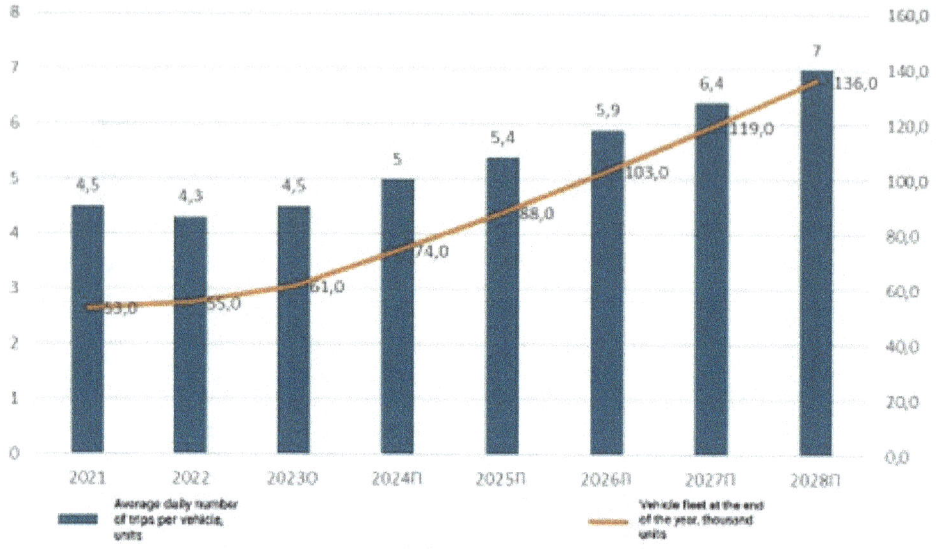

Based on the data presented in the figure above, the authors observe that the marginal decline in the potential of the car-sharing services market can be attributed to several challenges faced by companies, as previously discussed in the study. Moving forward, the market potential is projected to grow, both in terms of the number of trips per vehicle and the size of the car fleet.

Furthermore, the advancement of car-sharing aligns with one of the strategies7 aimed at reducing carbon dioxide emissions: the large-scale promotion of shared vehicle usage, alongside the development of bicycle and other non-motorized transport options. This approach emphasizes curbing the growth of private car ownership and improving quality of life through the expansion of shared consumption practices.

Figure 5. Dynamics of the share of the main participants in the carsharing services market in Russia8

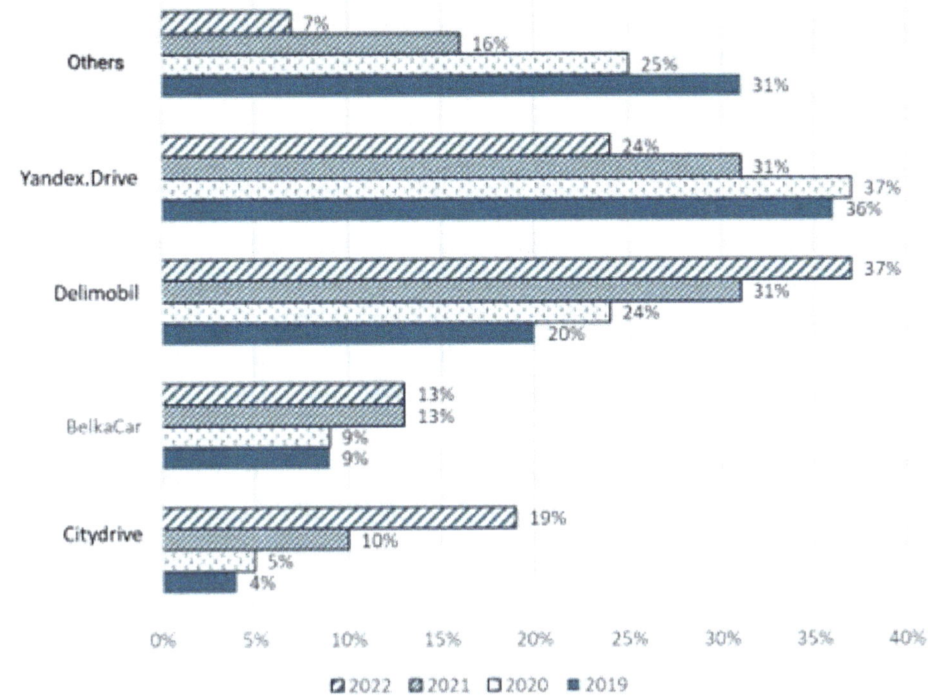

The main players in the market are Delimobil, Yandex.Drive, and Citydrive. Notably, Citydrive is strengthening its position in the market. In 2019, the operator held 4% of the market share, but by 2022, this figure had increased to 19%. Delimobil also expanded its market share, rising from 20% to 37% during the analyzed period. Meanwhile, Yandex.Drive, despite being among the top three players, experienced a decline in its market share in 2021 and 2022. In 2019, Yandex.Drive held 36% of the market, but by 2022, its share had decreased to 24%. Additionally, the combined market share of other companies has also been declining, reaching only 7% in 2022 (Research of the Shared Mobility and Carsharing Market, 2023).

CONCLUSION

Based on the analysis, the authors have drawn the following conclusions:

- The dominant position of the main market players persists, indicating that the competitive environment in the market has not improved.
- There is significant potential for the development of carsharing services in regional markets.
- There is a noticeable polarization in the development of the market across different territories.

In light of these findings, the authors propose the following measures:

- Development of regional carsharing markets by supporting regional operating companies and promoting these services among the local population.
- Enhancing the security of personal data for users of carsharing services.
- Improving territorial infrastructure to increase the attractiveness of operating companies' services in both the B2B and B2C markets.

The proposed measures are not exhaustive and can be further refined or expanded based on the specific characteristics of individual regions and emerging trends.

REFERENCES

Antonelli, C. (2003). The digital divide: Understanding the economics of new information and communication technologies in the global economy. *Information Economics and Policy*, *15*(2), 173–199. DOI: 10.1016/S0167-6245(02)00093-8

Baptista, P., Melo, S., & Rolim, C. (2014). Energy, environmental and mobility impacts of car-sharing systems: Empirical results from Lisbon, Portugal. *Procedia: Social and Behavioral Sciences*, *111*, 28–37. DOI: 10.1016/j.sbspro.2014.01.035

Carlsson, B. (2004). The digital economy: What's new and what's not? *Structural Change and Economic Dynamics*, *15*(3), 245–264. DOI: 10.1016/j.strueco.2004.02.001

Carmona, M., Heath, T., Tiesdell, S., & Oc, T. (2010). Urban design today. In *Public places, urban spaces: The dimensions of urban design* (2nd ed., pp. 3–19). Architectural Press., DOI: 10.1016/B978-1-85617-827-3.10001-X

Digilina, O. B., & Mironova, T. G. (2019). Перспективы развития каршеринга в России и за рубежом [Prospects for the development of car sharing in Russia and abroad]. Economics and Management: Problems. *Solutions*, *2*(3), 92–98.

Digital Transport and Logistics Association. (2023, December). Digital transformation of the transport and logistics industry of the Russian Federation: Trends, challenges, solutions, technologies. https://www.dtla.ru/upload/docs/Analitika_DTLA.pdf

Federal State Statistics Service. (n.d.). Federal State Statistics Service. Retrieved February 7, 2025, from http://rosstat.gov.ru/

Government of the Russian Federation. (2021, November 27). Transport strategy of the Russian Federation until 2030 with a forecast for the period until 2035 (Order No. 3363-r). SPS Consultant-Plus.

B1Group of Companies. (2023, December). Research of the shared mobility and carsharing market in Russia. https://b1.ru/analytics/b1-car-sharing-in-russia-survey/

Karlov, D. N., Polozhentseva, Y. S., Kremleva, L. V., & Kalimullin, D. D. (2019). Implementation of the IoT concept in the post-industrial economy. Espacios, 40(38), Article 28. https://www.revistaespacios.com/a19v40n38/19403828.html

Lammi, M., & Pantzar, M. (2019). The data economy: How technological change has shaped the citizen-consumer. *Technology in Society*, *59*, 101157. Advance online publication. DOI: 10.1016/j.techsoc.2019.101157

Luchko, M. I., Tsarev, M. N., & Kovtun, N. V. (2020). Анализ каршеринговых сервисов - преимущества и недостатки [Analysis of carsharing services: Advantages and disadvantages]. *Bulletin of the Volodymyr Dahl Luhansk National University*, (7), 128–131.

Nikanorova, L. V., & Loseva, M. V. (2019). Краткосрочная аренда автомобилей - каршеринг [Short-term car rental: Carsharing]. *Modern Technologies and Scientific and Technical Progress*, *1*, 198–199. DOI: 10.36629/2686-9896/2019-1-1-198-199

Podgaiskaya, A. E. (2019). Факторы успеха каршеринговых сервисов в России [Factors of success of carsharing services in Russia]. Region. *Economia e Sociologia*, (2), 270–289. DOI: 10.15372/REG20190212

Pyankova, S., & Ergunova, O. (2018). The role of regional development institutions in socio-economic development. In SGEM2018 Conference Proceedings: 5th International Multidisciplinary Scientific Conference on Social Sciences and Art (Vol. 5, Issue 1.3, pp. 885–892). SGEM. https://www.sgemsocial.org/ssgemlib/spip.php?article6063&lang=en

Semyachkov, K. A. (2025). Combined mechanism of public management of a smart city in the digitalization of ecosystem interactions (Doctoral dissertation). Yekaterinburg.

Susarenko, M. Y. (2023). Car sharing as an integral element of the development of the electric vehicle transport market. *The Scientific Heritage*, (122), 17–19. DOI: 10.5281/zenodo.8412786

Unified Interdepartmental Information and Statistical System. (n.d.). Unified Interdepartmental Information and Statistical System. Retrieved February 7, 2025, from https://www.fedstat.ru/indicator/58939

Compilation of References

Abam, F. I., Ekwe, B. E., Diemuodeke, O. E., Ofem, M. I., Okon, B. B., Kadurumba, C. H., Archibong-Eso, A., Effiom, S. O., Egbe, J. G., & Ukueje, W. E. (2021). Environmental sustainability of the Nigerian transport sector through decomposition and decoupling analysis with future framework for sustainable transport pathways. *Energy Reports*, *7*, 3238–3248. DOI: 10.1016/j.egyr.2021.05.044

Abedin, M., Khan, S., & Hasan, M. A. (2021). A deep learning-based framework for automated road anomaly detection using UAV imagery. *IEEE Access : Practical Innovations, Open Solutions*, *9*, 66545–66556. DOI: 10.1109/ACCESS.2021.3075745

Ahmad, A., Alam, M. S., & Chabaan, R. (2018). A comprehensive review of wireless charging technologies for electric vehicles. *IEEE Transactions on Transportation Electrification*, *4*(1), 38–63. DOI: 10.1109/TTE.2017.2771619

Ahmad, A., Khalid, M., Ullah, Z., Ahmad, N., Aljaidi, M., Malik, F. A., & Manzoor, U. (2022). Electric vehicle charging modes, technologies and applications of smart charging. *Energies*, *15*(24), 9471. Advance online publication. DOI: 10.3390/en15249471

Akbulaev, N., & Bayramli, G. (2020). Maritime transport and economic growth: Interconnection and influence (an example of the countriesin the Caspian sea coast; Russia, Azerbaijan, Turkmenistan, Kazakhstan and Iran). *Marine Policy*, *118*, 104005. DOI: 10.1016/j.marpol.2020.104005

Aktar, A. K., Taşcıkaraoğlu, A., Erdinç, O., & Güner, S. (2024). Impacts of Distribution-Level Joint Scheduling of Electric Vehicle Battery Charging and Swapping Stations on Reliability Improvement. IEEE Transactions on Industry Applications. https://ieeexplore.ieee.org/abstract/document/10561522

Alanazi, F. (2023). Electric vehicles: Benefits, challenges, and potential solutions for widespread adaptation. *Applied Sciences (Basel, Switzerland)*, *13*(10), 6016. Advance online publication. DOI: 10.3390/app13106016

Alhaddi, H. (2015). Triple bottom line and sustainability: A literature review. *Business and Management Studies*, *1*(2), 6–10. DOI: 10.11114/bms.v1i2.752

Ali, H., Hussain, S., Khan, H. A., Arshad, N., & Khan, I. A. (2020, September). Economic and environmental impact of vehicle-to-grid (V2G) integration in an intermittent utility grid. In 2020 2nd International Conference on Smart Power & Internet Energy Systems (SPIES) (pp. 345-349). IEEE. https://www.mdpi.com/2071-1050/14/21/13856

Al-karakchi, A. A. A., Lacey, G., & Putrus, G. (2015, September). A method of electric vehicle charging to improve battery life. In 2015 50th International Universities Power Engineering Conference (UPEC) (pp. 1-3). IEEE. https://ieeexplore.ieee.org/abstract/document/7339846

Al-Sahlawi, A. A. K., Ayob, S. M., Tan, C. W., Ridha, H. M., & Hachim, D. M. (2024). Optimal Design of Grid-Connected Hybrid Renewable Energy System Considering Electric Vehicle Station Using Improved Multi-Objective Optimization: Techno-Economic Perspectives. *Sustainability (Basel)*, *16*(6), 2491. https://www.mdpi.com/2071-1050/16/6/2491. DOI: 10.3390/su16062491

Anegawa, T. (2010). Development of quick charging system for electric vehicle. https://www.osti.gov/etdeweb/biblio/21423136

Antonelli, C. (2003). The digital divide: Understanding the economics of new information and communication technologies in the global economy. *Information Economics and Policy*, *15*(2), 173–199. DOI: 10.1016/S0167-6245(02)00093-8

Astratova., ... (2024). Interaction of the system of higher professional education, R&D and the labor market in the context of digitalization of the economy of the Russian Federation and the Republic of Belarus: a collective scientific monograph. - Yekaterinburg: Azhur Publishing House, 2024. – 274 pp. URL: https://elar.urfu.ru/handle/10995/139939 (In Russ.).

Astratova, G. V., Rutkauskas, T. K., Rutkauskas, K. V., & Klimuk, V. V. (2021). Creating an environmentally safe and reliable heat supply system through the introduction of energy-saving technologies in housing and common utilities services. In *E3S web of conferences* (Vol. 265, p. 04020). EDP Sciences., DOI: 10.1051/e3sconf/202126504020

Azad, A. N., Echols, A., Kulyukin, V. A., Zane, R., & Pantic, Z. (2019). Analysis, optimization, and demonstration of a vehicular detection system intended for dynamic wireless charging applications. *IEEE Transactions on Transportation Electrification*, *5*(1), 147–161. DOI: 10.1109/TTE.2018.2870339

Azar, A., & Kumar, R. (2020). Integration of GIS with remote sensing for urban road damage mapping. *International Journal of Geoinformatics, 16*(3), 42–52.

B1Group of Companies. (2023, December). Research of the shared mobility and carsharing market in Russia. https://b1.ru/analytics/b1-car-sharing-in-russia-survey/

Balali, V., Golroo, A., & Law, K. H. (2021). A vision-based system for pothole detection using deep learning. *Transportation Research Record: Journal of the Transportation Research Board, 2675*(3), 222–231. DOI: 10.1177/0361198120987642

Baptista, P., Melo, S., & Rolim, C. (2014). Energy, environmental and mobility impacts of car-sharing systems: Empirical results from Lisbon, Portugal. *Procedia: Social and Behavioral Sciences, 111*, 28–37. DOI: 10.1016/j.sbspro.2014.01.035

Barns, S., Cosgrave, E., Acuto, M., & McNeill, D. (2016). Digital infrastructures and urban governance. *Urban Policy and Research, 35*(1), 20–31. DOI: 10.1080/08111146.2016.1235032

Bastani, F., & Hassani, S. (2019). Road maintenance optimization using geospatial tools and machine learning techniques. *Journal of Infrastructure Systems, 25*(4), 04019022. DOI: 10.1061/(ASCE)IS.1943-555X.0000515

Bayram, I. S., & Tajer, A. 2017. Plug-in electric vehicle grid integration. Artech House. https://pureportal.strath.ac.uk/en/publications/plug-in-electric-vehicle-grid-integration

Berliner Verkehrsbetriebe. (n.d.). Connection search. https://www.bvg.de/en/connections/connection-search

Biswas, A., & Wang, H. C. (2023). Autonomous vehicles enabled by the integration of IoT, edge intelligence, 5G, and blockchain. *Sensors, 23*(4), 1963. https://www.mdpi.com/1424-8220/23/4/1963# Bösch, P. M., Ciari, F., &Axhausen, K. (2018). Transport policy optimization with autonomous vehicles. Transportation Research Record, 2672(8), 698-707. https://doi.org/DOI: 10.1177/0361198118791391

Bosshard, R., Mühlethaler, J. W., Kolar, J. W., & Stevanović, I. (2013). Optimized magnetic design for inductive power transfer coils. In *2013 Twenty-Eighth Annual IEEE Applied Power Electronics Conference and Exposition (APEC)* (pp. 1812–1819). IEEE. DOI: 10.1109/APEC.2013.6520541

Botsford, C., & Szczepanek, A. (2009, May). Fast charging vs. slow charging: Pros and cons for the new age of Electric Vehicles. In *International Battery Hybrid Fuel Cell Electric Vehicle Symposium* (pp. 1–9). Citeseer., https://static1.squarespace.com/static/55f4e4d1e4b029b54a77cf2d/t/55fdf7a9e4b016bd35401ea1/1442707369493/EVS24_3960315_Botsford_r.pdf

Boys, J. T., Covic, G. A., & Green, A. W. (2000). Stability and control of inductively coupled power transfer systems. *IEE Proceedings. Electric Power Applications*, *147*(1), 37–43. https://digital-library.theiet.org/content/journals/10.1049/ip-epa_20000017. DOI: 10.1049/ip-epa:20000017

Braun, V., & Clarke, V. (2006). Using thematic analysis in psychology. *Qualitative Research in Psychology*, *3*(2), 77–101. DOI: 10.1191/1478088706qp063oa

Brown, A., Cappellucci, J., Heinrich, A., & Cost, E. (2024). Electric vehicle charging infrastructure trends from the alternative fueling station locator: Third quarter 2023 (No. NREL/TP-5400-88223). National Renewable Energy Laboratory (NREL), Golden, CO (United States). https://www.osti.gov/biblio/2293493

Bucher, A., & Duerbaum, T. (2012). Analysis and design of a contactless power transmission system based on the extended first harmonic approximation. In *2012 IEEE Industry Applications Society Annual Meeting* (pp. 1–7). IEEE. DOI: 10.1109/IAS.2012.6374114

Buchwald, E., & Valentik, O. (2020). Municipal strategies in Russia: New conditions and new challenges. Regional Economy. *South of Russia*, *8*(1), 33–43. DOI: 10.15688/re.volsu.2020.1.3

Budhia, M., Covic, G. A., Boys, J. T., & Huang, C. Y. (2011, September). Development and evaluation of single sided flux couplers for contactless electric vehicle charging. In 2011 IEEE Energy Conversion Congress and Exposition (pp. 614-621). IEEE. https://ieeexplore.ieee.org/abstract/document/6063826

Budhia, M., Boys, J. T., Covic, G. A., & Huang, C.-Y. (2011). Development of a single-sided flux magnetic coupler for electric vehicle IPT charging systems. *IEEE Transactions on Industrial Electronics*, *60*(1), 318–328. DOI: 10.1109/TIE.2011.2179274

Budhia, M., Covic, G. A., & Boys, J. T. (2011). Design and optimization of circular magnetic structures for lumped inductive power transfer systems. *IEEE Transactions on Power Electronics*, *26*(11), 3096–3108. DOI: 10.1109/TPEL.2011.2143730

Butler, L., Yigitcanlar, T., & Paz, A. (2020). Smart urban mobility innovations: A comprehensive review and evaluation. *IEEE Access : Practical Innovations, Open Solutions*, *8*, 196034–196049. DOI: 10.1109/ACCESS.2020.3034596

Camagni, R., Gibelli, M. C., & Rigamonti, P. (2002). Urban mobility and urban form: The social and environmental costs of different patterns of urban expansion. *Ecological Economics*, *40*(2), 199–216. DOI: 10.1016/S0921-8009(01)00254-3

Can, E. (2024). DC–DC converter with multiple inputs and full isolated multi ports charging battery in photovoltaic energy systems. *Sadhana*, *49*(2), 148. https://link.springer.com/article/10.1007/s12046-024-02506-y. DOI: 10.1007/s12046-024-02506-y

Carlsson, B. (2004). The digital economy: What's new and what's not? *Structural Change and Economic Dynamics*, *15*(3), 245–264. DOI: 10.1016/j.strueco.2004.02.001

Carmona, M., Heath, T., Tiesdell, S., & Oc, T. (2010). Urban design today. In *Public places, urban spaces: The dimensions of urban design* (2nd ed., pp. 3–19). Architectural Press., DOI: 10.1016/B978-1-85617-827-3.10001-X

Ceder, A. (2020). Urban mobility and public transport: Future perspectives and review. *International Journal of Urban Sciences*, *25*(4), 455–479. DOI: 10.1080/12265934.2020.1799846

Chakraborty, P., & Pal, M. (2024). Planning of fast charging infrastructure for Electric Vehicles in a distribution system and prediction of dynamic price. *International Journal of Electrical Power & Energy Systems*, *155*, 109502. https://www.sciencedirect.com/science/article/pii/S0142061523005598. DOI: 10.1016/j.ijepes.2023.109502

Chandrakar, N., Sekhar, C. C., Mageshwari, S., & Daniel, S. A. (2024). Electric vehicle progression in the society and their consequences. In *Advanced Technologies in Electric Vehicles* (pp. 127–160). Academic Press., https://www.sciencedirect.com/science/article/abs/pii/B9780443189999000053 DOI: 10.1016/B978-0-443-18999-9.00005-3

Chen, Y., Wang, Z., & Ortiz, J. A. (2023). A sustainable digital ecosystem: Digital servitization transformation and digital infrastructure support. *Sustainability (Basel)*, *15*(2), 1530. DOI: 10.3390/su15021530

Chigbu, B. I. (2024). Advancing sustainable development through circular economy and skill development in EV lithium-ion battery recycling: A comprehensive review. *Frontiers in Sustainability*, *5*, 1409498. https://www.frontiersin.org/journals/sustainability/articles/10.3389/frsus.2024.1409498/full. DOI: 10.3389/frsus.2024.1409498

Chirisa, I., Bandauko, E., & Gaza, M. (2015). Transport and the Environment: A Critical Review of Africa. *Zhongguo Renkou Ziyuan Yu Huanjing*, *13*(4), 309–319. Advance online publication. DOI: 10.1080/10042857.2015.1078489

Chowdary, K. V., Kumar, K., Behera, R. K., & Banerjee, S. (2020). Overview and analysis of various coil structures for dynamic wireless charging of electric vehicles. In *2020 IEEE International Conference on Power Electronics, Smart Grid and Renewable Energy (PESGRE2020)* (pp. 1–6). IEEE. DOI: 10.1109/PESGRE45664.2020.9070496

Chowdhury, M. S. A., & Liang, X. (2019a). Comparative characteristic analysis of circular and double D power pads for electric vehicle wireless charging systems. In *2019 IEEE Canadian Conference of Electrical and Computer Engineering (CCECE)* (pp. 1–4). IEEE. DOI: 10.1109/CCECE.2019.8861733

Chowdhury, M. S. A., & Liang, X. (2019b). Power transfer efficiency evaluation of different power pads for electric vehicle's wireless charging systems. In *2019 IEEE Canadian Conference of Electrical and Computer Engineering (CCECE)* (pp. 1–4). IEEE. DOI: 10.1109/CCECE.2019.8861757

Costa, P. B., Morais Neto, G. C., & Bertolde, A. I. (2017). Urban mobility indexes: A brief review of the literature. *Transportation Research Procedia*, *25*, 3645–3655. DOI: 10.1016/j.trpro.2017.05.330

Cota, K. A., Gray, P. A., Pathmanathan, M., & Lehn, P. W. (2019). An approach for selecting compensation capacitances in resonance-based EV wireless power transfer systems with switched capacitors. *IEEE Transactions on Transportation Electrification*, *5*(4), 1004–1014. DOI: 10.1109/TTE.2019.2927803

Covic, G. A., & Boys, J. T. (2013). Modern trends in inductive power transfer for transportation applications. *IEEE Journal of Emerging and Selected Topics in Power Electronics*, *1*(1), 28–41. DOI: 10.1109/JESTPE.2013.2264473

Covic, G. A., Boys, J. T., Budhia, M., & Huang, C.-Y. (2010). Electric vehicles–personal transportation for the future. *World Electric Vehicle Journal*, *4*(4), 693–704. DOI: 10.3390/wevj4040693

Creswell, J. W., & Poth, C. N. (2018). *Qualitative Inquiry and Research Design: Choosing Among Five Approaches*. Sage Publications.

Dai, X., Li, X., Li, Y., & Hu, A. P. (2018). Maximum efficiency tracking for wireless power transfer systems with dynamic coupling coefficient estimation. *IEEE Transactions on Power Electronics*, *33*(6), 5005–5015. DOI: 10.1109/TPEL.2017.2729083

Daly, H. E. (1991). *Steady State Economics*. Island Press.

Dawkins, E., André, K., Leander, E., Axelsson, K., & Gerger Swartling, Å. (2023). Policy for sustainable consumption — An assessment of Swedish municipalities. *Frontiers in Sustainability*, *4*, 1265733. DOI: 10.3389/frsus.2023.1265733

Dementiev, A. (2016). Strategic partnerships in local public transport. *Research in Transportation Economics*, *59*, 65–74. DOI: 10.1016/j.retrec.2016.07.017

Dementiev, A., & Loboyko, A. (2014). Trusting partnerships in a regulatory game: The case of suburban railway transport in Russia. *Research in Transportation Economics*, *48*, 209–220. DOI: 10.1016/j.retrec.2014.09.045

Deng, J., Lu, F., Li, S., Nguyen, T.-D., & Mi, C. (2014). Development of a high efficiency primary side controlled 7kW wireless power charger. In *2014 IEEE International Electric Vehicle Conference (IEVC)* (pp. 1–6). IEEE. DOI: 10.1109/IEVC.2014.7056204

Denzin, N. K., & Lincoln, Y. S. (2018). *The Sage Handbook of Qualitative Research*. Sage Publications.

Diep, N. T., Trung, N. K., & Minh, T. T. (2019). Maximum efficiency in the dynamic wireless charging systems of electric vehicles. In 2019 10th International Conference on Power Electronics and ECCE Asia (ICPE 2019—ECCE Asia) (pp. 1–6). IEEE. DOI: 10.23919/ICPE2019-ECCEAsia42246.2019.8797184

Diep, N. T., Trung, N. K., & Minh, T. T. (2020). Wireless power transfer system design for electric vehicle dynamic charging application. *International Journal of Power Electronics and Drive Systems*, *11*(3), 1468–1480. DOI: 10.11591/ijpeds.v11.i3.pp1468-1480

Digilina, O. B., & Mironova, T. G. (2019). Перспективы развития каршеринга в России и за рубежом [Prospects for the development of car sharing in Russia and abroad]. Economics and Management: Problems. *Solutions*, *2*(3), 92–98.

Digital Transport and Logistics Association. (2023, December). Digital transformation of the transport and logistics industry of the Russian Federation: Trends, challenges, solutions, technologies. https://www.dtla.ru/upload/docs/Analitika_DTLA.pdf

Dirgahayani, P. (2013). Environmental Co-benefits of Public Transportation Improvement Initiative: The Case of Trans-Jogja Bus System in Yogyakarta, Indonesia. *Journal of Cleaner Production*, *58*, 74–81. DOI: 10.1016/j.jclepro.2013.07.013

Dopravní podnik hlavního města Prahy. (n.d.). Ceník jízdného [Fare prices]. https://www.dpp.cz/jizdne/cenik-jizdneho

Dovlatova, G., Zhukova, I., Bondarenko, O., Agafonov, A., &Zheltushkina, E. (2023). International analytics of the transport sector of the economy and its adaptation to Russian conditions. E3S Web of Conferences, 371, 05033. https://doi.org/DOI: 10.1051/e3sconf/202337105033

Dubey, A., Singh, P., & Tripathi, R. K. (2020). Automated road anomaly detection using mobile sensors and CNN-based algorithms. *Sensors (Basel)*, *20*(6), 1755. DOI: 10.3390/s20061755 PMID: 32245278

Dumas, B. L., Harris, D. M., McMahon, J., Daymude, T. J., Warnock, A. L., Moore, L., & Onufrak, S. J. (2021). Prevalence of municipal-level policies dedicated to transportation that consider food access. *Preventing Chronic Disease*, *18*, 210193. DOI: 10.5888/pcd18.210193 PMID: 34793691

Duranton, G., & Turner, M. A. (2011). The fundamental law of road congestion: Evidence from US cities. *The American Economic Review*, *101*(6), 2616–2652. DOI: 10.1257/aer.101.6.2616

Du, Y., Lukic, S., Jacobson, B., & Huang, A. (2011, September). *Review of high power isolated bi-directional DC-DC converters for PHEV/EV DC charging infrastructure. In 2011 IEEE Energy Conversion Congress and Exposition*. IEEE., https://ieeexplore.ieee.org/abstract/document/6063818

Du, Z.-Y., & Wang, Q. (2024). Digital infrastructure and innovation: Digital divide or digital dividend? *Journal of Innovation & Knowledge*, *9*(3), 100542. DOI: 10.1016/j.jik.2024.100542

Ebzeev, A. A., Sudorgin, O., Nesterchuk, O., &Ivanov, V. G. (2023). The impact of politics on transport system of the Russian Federation. UPRAVLENIE / MANAGEMENT (Russia), 11(2), 35-44. https://doi.org/DOI: 10.26425/2309-3633-2023-11-2-35-44

Ehsani, M., Gao, Y., & Emadi, A. (2010). *Modern electric, hybrid electric, and fuel cell vehicles: Fundamentals, theory, and design* (2nd ed.). CRC Press., DOI: 10.1201/9781420054002

Elkington, J. (1997). *Cannibals with forks – Triple bottom line of 21st-century business*. New Society Publishers.

Elkington, J. (1997). *Cannibals with Forks: The Triple Bottom Line of 21st Century Business*. Capstone Publishing.

Emadi, A., Ehsani, M., & Miller, J. M. (2003). *Vehicular electric power systems: Land, sea, air, and space vehicles*. CRC Press., DOI: 10.1201/9780203913468

English, R. (2024). Raising Public Awareness of Electric Vehicle Technology to Increase Sales. https://scholarworks.waldenu.edu/dissertations/15388/

ESCAP. U. (2024). Accelerating electric mobility in Public Transport: how Asia and the Pacific could unlock financing for electric buses. https://repository.unescap.org/handle/20.500.12870/6756

Esposito, F., & Naumann, D. (2021). Machine vision for infrastructure inspection: Applications and challenges. *IEEE Transactions on Industrial Informatics*, *17*(6), 3912–3925. DOI: 10.1109/TII.2020.3018983

Etacheri, V., Marom, R., Elazari, R., Salitra, G., & Aurbach, D. (2011). Challenges in the development of advanced Li-ion batteries: A review. *Energy & Environmental Science*, *4*(9), 3243–3262. DOI: 10.1039/c1ee01598b

Fasugba, M. A., & Krein, P. T. (2011, September). Gaining vehicle-to-grid benefits with unidirectional electric and plug-in hybrid vehicle chargers. In 2011 IEEE vehicle power and propulsion conference (pp. 1-6). IEEE. https://ieeexplore.ieee.org/abstract/document/6043207

Federal State Statistics Service. (n.d.). Federal State Statistics Service. Retrieved February 7, 2025, from http://rosstat.gov.ru/

Feng, Y., & Lu, X. (2021). Construction planning and operation of battery swapping stations for electric vehicles: A literature review. *Energies*, *14*(24), 8202. Advance online publication. DOI: 10.3390/en14248202

Filimonau, V., & Ermolaev, V. (2021). A sleeping giant? Food waste in the food-service sector of Russia. *Journal of Cleaner Production*, *126705*, 126705. Advance online publication. DOI: 10.1016/j.jclepro.2021.126705

Foltýnová, H. B., Vejchodská, E., Rybová, K., & Květoň, V. (2020). Sustainable urban mobility: One definition, different stakeholders' opinions. *Transportation Research Part D, Transport and Environment*, *87*, 102465. DOI: 10.1016/j.trd.2020.102465

Fotouhi, A., Auger, D. J., Propp, K., Longo, S., & Wild, M. (2016). A review on electric vehicle battery modelling: From Lithium-ion toward Lithium–Sulphur. *Renewable & Sustainable Energy Reviews*, *56*, 1008–1021. https://www.sciencedirect.com/science/article/abs/pii/S1364032115013921. DOI: 10.1016/j.rser.2015.12.009

Freemark, Y., Hudson, A., & Zhao, J. (2019). Are cities prepared for autonomous vehicles? *Journal of the American Planning Association*, *85*(2), 133–151. DOI: 10.1080/01944363.2019.1603760

Freemark, Y., Hudson, A., & Zhao, J. (2020). Policies for autonomy: How American cities envision regulating automated vehicles. *Urban Science (Basel, Switzerland)*, *4*(4), 55. DOI: 10.3390/urbansci4040055

Gakenheimer, R. (1999). Urban mobility in the developing world. *Transportation Research Part A, Policy and Practice*, *33*(7–8), 671–689. DOI: 10.1016/S0965-8564(99)00005-1

Ganbold, U. (2024). Municipal-private partnership in Russia and Mongolia. *Public Administration*, *26*(2), 78–82. DOI: 10.22394/2070-8378-2024-26-2-78-82

Gao, Y., Duan, C., Oliveira, A. A., Ginart, A., Farley, K. B., & Tse, Z. T. H. (2017). 3-D coil positioning based on magnetic sensing for wireless EV charging. *IEEE Transactions on Transportation Electrification*, *3*(3), 578–588. DOI: 10.1109/TTE.2017.2696787

Godnev, A. (2020). Intercepted parking as a method of combating transport convenience of the megapolis. salogistics.ru. http://www.salogistics.ru/magazine/26/15_godnev_115-121.pdf

Goel, P. (2010). Triple bottom line reporting: An analytical approach for corporate sustainability. *Journal of Finance, Accounting, and Management*, *1*(1), 27–42.

Gordon, L. B., Carr, K. D., & Graham, N. (2017). Complete electrical arc hazard classification system and its application. *IEEE Transactions on Industry Applications*, *53*(5), 5078–5087. https://ieeexplore.ieee.org/abstract/document/7878523/. DOI: 10.1109/TIA.2017.2682158

Goryachenko, E. E., & Deineko, L. (2015). Problems of municipal management in Russia and Ukraine. *Regional Research of Russia*, *5*(4), 375–382. DOI: 10.1134/S2079970515030053

Government of the Russian Federation. (2021, November 27). Transport strategy of the Russian Federation until 2030 with a forecast for the period until 2035 (Order No. 3363-r). SPS Consultant-Plus.

Grand View Research. (n.d.). *Smart cities market size, share & trends analysis report by application, by smart utilities, by smart transportation, by smart healthcare, 2024–2030: Market analysis report*. Grand View Research, Inc.

Greenstein, S. (2021). Digital infrastructure. In Glaeser, E. L., & Poterba, J. M. (Eds.), *Economic analysis and infrastructure investment* (pp. 409–452). University of Chicago Press., DOI: 10.7208/chicago/9780226800615.003.0009

Guo, B., Hu, P., & Lin, J. (2024). The effect of digital infrastructure development on enterprise green transformation. *International Review of Financial Analysis*, *92*, 103085. DOI: 10.1016/j.irfa.2024.103085

Gupta, R., Gupta, D., & Sharma, K. (2021). GIS-based pothole management systems: A case study. *Journal of Geographic Information System, 13*(1), 67–82. DOI: 10.4236/jgis.2021.131004

Gurtu, A. (2014). A Framework for Reducing Greenhouse Gas (GHG) Emissions through Carbon Pricing for Offshore Outsourcing. (Doctoral dissertation, Ryerson University, Toronto). Retrieved from http://digital.library.ryerson.ca/islandora/object/RULA%3A3388

Guzek, M., Jackowski, J., Jurecki, R. S., Szumska, E. M., Zdanowicz, P., & Żmuda, M. (2024). Electric Vehicles—An Overview of Current Issues—Part 2—Infrastructure and Road Safety. *Energies, 17*(2), 495. https://www.mdpi.com/1996-1073/17/2/495. DOI: 10.3390/en17020495

Gvozdkov, Y. Y. (2019). As an example of a sustainable economic policy of a municipal government in Czarist Russia. *Economy and History, 15*(2), 176–187. DOI: 10.15507/2409-630X.045.015.201902.176-187

Hardman, S., Shiu, E., & Steinberger-Wilckens, R. (2016). Comparing high-end and low-end early adopters of battery Electric Vehicles. *Transportation Research Part A, Policy and Practice, 88*, 40–57. https://www.sciencedirect.com/science/article/pii/S0965856416302208. DOI: 10.1016/j.tra.2016.03.010

Harper, J. D. (2013). Development and implementation of sae dc charging digital communication for plug-in electric vehicle dc charging (No. 2013-01-1188). SAE Technical Paper. https://www.sae.org/publications/technical-papers/content/2013-01-1188/

Hata, K., Imura, T., & Hori, Y. (2016). *Dynamic wireless power transfer system for electric vehicles to simplify ground facilities—power control and efficiency maximization on the secondary side. In 2016 IEEE Applied Power Electronics Conference and Exposition (APEC).* IEEE., DOI: 10.1109/APEC.2016.7468105

Henfridsson, O., & Bygstad, B. (2013). The generative mechanisms of digital infrastructure evolution. *Management Information Systems Quarterly, 37*(3), 907–931. https://www.jstor.org/stable/43826006. DOI: 10.25300/MISQ/2013/37.3.11

Henriques, A. (2007). CSR, sustainability and the triple bottom line. In Henriques, A., & Richardson, J. (Eds.), *The Triple Bottom Line: Does it All Add Up?* (pp. 26–33). Earthscan.

Hill, C. (2009). Transportation and sustainability best practices background. *Proceedings, AASHTO Sustainability Peer Exchange*, Washington, DC, 32.

Hossain, A., Darvish, P., Mekhilef, S., Tey, K. S., & Tong, C. W. (2022). A new coil structure of dual transmitters and dual receivers with integrated decoupling coils for increasing power transfer and misalignment tolerance of wireless EV charging system. *IEEE Transactions on Industrial Electronics*, *69*(8), 7869–7878. DOI: 10.1109/TIE.2021.3108697

Ibrahim, M., Pichon, L., Bernard, L., Razek, A., Houivet, J., & Cayol, O. (2015). Advanced modeling of a 2-kW series–series resonating inductive charger for real electric vehicle. *IEEE Transactions on Vehicular Technology*, *64*(2), 421–430. DOI: 10.1109/TVT.2014.2325614

IEA. *Energy Policies of IEA Countries: Greece 2011 Review*. OECD Pub., 2011. https://www.iea.org/reports/energy-policies-of-iea-countries-greece-2011-review

Interfax. (2011, December 2). В Москве утверждены новые тарифы на проезд в общественном транспорте [New public transport fares approved in Moscow]. https://www.interfax.ru/russia/217432

International Energy Agency. Global EV Outlook 2018—Towards Cross-modal Electrification; International Energy Agency: Paris, France. https://www.iea.org/reports/global-ev-outlook-2018

Jain, F., & Bhullar, S. (2024). Operating Modes of Grid Integrated PV-Solar Based Electric Vehicle Charging System-A Comprehensive Review. e-Prime-Advances in Electrical Engineering, Electronics and Energy, 100519. https://www.sciencedirect.com/science/article/pii/S2772671124001013

Jamakani, E., Afjei, E., & Mosallanejad, A. (2019). A novel triple quadrature pad for inductive power transfer systems for electric vehicle charging. In 2019 10th International Power Electronics, Drive Systems and Technologies Conference (PEDSTC) (pp. 618–623). IEEE. DOI: 10.1109/PEDSTC.2019.8697654

Jan, B., Farman, H., Khan, M., Talha, M., & Din, I. U. (2019). Designing a smart transportation system: An internet of things and big data approach. *IEEE Wireless Communications*, *26*(4), 73–79. DOI: 10.1109/MWC.2019.1800512

Jeong, S. Y., Kwak, H. G., Jang, G. C., Choi, S. Y., & Rim, C. T. (2018). Dual-purpose nonoverlapping coil sets as metal object and vehicle position detections for wireless stationary EV chargers. *IEEE Transactions on Power Electronics*, *33*(9), 7387–7397. DOI: 10.1109/TPEL.2017.2765521

Jimenez, J. A. (2018). Smart transportation systems. In McClellan, S., Jimenez, J., & Koutitas, G. (Eds.), *Smart cities* (pp. 143–160). Springer., DOI: 10.1007/978-3-319-59381-4_8

Jin, X., & Meintz, A. (2015). Challenges and Opportunities for Transactive Control of Electric Vehicle Supply Equipment. A Reference Guide. https://www.osti.gov/biblio/1215194

Jiwariyavej, V., Imura, T., & Hori, Y. (2015). Coupling coefficients estimation of wireless power transfer system via magnetic resonance coupling using information from either side of the system. *IEEE Journal of Emerging and Selected Topics in Power Electronics*, *3*(1), 191–200. DOI: 10.1109/JESTPE.2014.2332056

JLL. (2017). *Residential city profile: Berlin*. https://jll-global-gdim-res.cloudinary.com/image/upload/v1524021097/olmwebcontents/ip/prod/mmlimougy5wnrbmzhrom.pdf

Kalwar, K. A., Mekhilef, S., Seyedmahmoudian, M., & Horan, B. (2016). Coil design for high misalignment tolerant inductive power transfer system for EV charging. *Energies*, *9*(11), 937. Advance online publication. DOI: 10.3390/en9110937

Kang, T., Chae, B., & Suh, Y. (2013, June). Control algorithm of bi-directional power flow rapid charging system for electric vehicle using Li-Ion polymer battery. In 2013 IEEE ECCE Asia Downunder (pp. 499-505). IEEE. https://ieeexplore.ieee.org/abstract/document/6579143

Kan, T., Lu, F., Nguyen, T.-D., Mercier, P. P., & Mi, C. C. (2018). Integrated coil design for EV wireless charging systems using LCC compensation topology. *IEEE Transactions on Power Electronics*, *33*(11), 9231–9241. DOI: 10.1109/TPEL.2018.2794448

Karami, Z., & Kashef, R. (2020). Smart transportation planning: Data, models, and algorithms. *Transportation Engineering*, *2*, 100013. DOI: 10.1016/j.treng.2020.100013

Karamov, D. (2020). Autonomous renewable energy systems in Russia. Critical review of the current situation. *Energy Reports*, *6*, 1584–1593. DOI: 10.1016/j.egyr.2020.10.033

Karlov, D. N., Polozhentseva, Y. S., Kremleva, L. V., & Kalimullin, D. D. (2019). Implementation of the IoT concept in the post-industrial economy. Espacios, 40(38), Article 28. https://www.revistaespacios.com/a19v40n38/19403828.html

Khalid, M., Thakur, J., Bhagavathy, S. M., & Topel, M. (2024). Impact of public and residential smart EV charging on distribution power grid equipped with storage. *Sustainable Cities and Society*, *104*, 105272. https://www.sciencedirect.com/science/article/pii/S221067072400101X. DOI: 10.1016/j.scs.2024.105272

Kim, S., Covic, G. A., & Boys, J. T. (2016a). Analysis on tripolar pad for inductive power transfer systems. In *2016 IEEE PELS Workshop on Emerging Technologies: Wireless Power Transfer (WoW)* (pp. 15–20). IEEE. DOI: 10.1109/WoW.2016.7772059

Kim, S., Covic, G. A., & Boys, J. T. (2016b). Tripolar pad for inductive power transfer systems for EV charging. *IEEE Transactions on Power Electronics*, *32*(7), 5045–5057. DOI: 10.1109/TPEL.2016.2606893

Klontz, K. W., Divan, D. M., Novotny, D. W., & Lorenz, R. D. (1994). *U.S. Patent No. 5,341,083*. U.S. Patent and Trademark Office., https://patents.google.com/patent/US5341083A/

Knoflacher, H. (2017). Verkehr ist kein Schicksal: Der öffentliche Verkehr in Wien [Traffic is not fate: Public transport in Vienna]. TU Wien. https://www.tuwien.at/index.php?eID=dumpFile&f=138722&t=f&token=b57c5dd11329b20318241ed0048d2c66474a7018

Kobayashi, K., Imura, T., & Hori, Y. (2015). Real-time coupling coefficient estimation and maximum efficiency control on dynamic wireless power transfer for electric vehicles. In *2015 IEEE PELS Workshop on Emerging Technologies: Wireless Power (WoW)* (pp. 1–6). IEEE. DOI: 10.1109/WoW.2015.7132799

Komarov., ... (2019). Komarov, Yu. Ya.,Ganzin, S. V., Komarov, D. Yu. The role of municipalities in improving the functioning of transport infrastructure. *Transport of the Russian Federation. A journal about science, practice, and economics*, 2 (81). URL: https://cyberleninka.ru/article/n/rol-munitsipalitetov-v-uluchshenii-funktsionirovaniya-transportnoy-infrastruktury (date of request: 12/21/2024). (In Russ.).

Koutsikouri, D., Lindgren, R., Henfridsson, O., & Rudmark, D. (2018). Extending digital infrastructures: A typology of growth tactics. *Journal of the Association for Information Systems*, *19*(10), 1001–1019. DOI: 10.17705/1jais.00517

Kovin, V., & Semenov, A. (2022). Political logic of the "Municipal Filter" in Russia: Evidence from Perm Krai 2020 elections. *Russian Politics*, *7*(2), 209–236. DOI: 10.30965/24518921-00604029

Kumar, R., Mahdu, E., Dahiya, A., & Sinha, S. (2015). Analytical Hierarchy Process for Assessing Sustainability: Indicators of Public Transportation Systems, Pedestrians and Feeder Services in Developing Countries. *World Journal of Science, Technology and Sustainable Development,* 12*(4).

Kumar, M., Panda, K. P., Rosas-Caro, J. C., Valderrabano-Gonzalez, A., & Panda, G. (2023). Comprehensive review of conventional and emerging maximum power point tracking algorithms for uniformly and partially shaded solar photovoltaic systems. *IEEE Access : Practical Innovations, Open Solutions, 11*, 31778–31812. https://ieeexplore.ieee.org/abstract/document/10083133. DOI: 10.1109/ACCESS.2023.3262502

Küpper, K., Pels, T., Deiml, M., Angermaier, A., Bürger, T., Weinzerl, A., & Teuschl, G. (2015). Tension 12 V to 800 V efficient powertrain solutions. *Efficient Powertrain Solutions for 12V Up to 800V*.

Kuznetsova, O. (2021). National urban policy in Russia and the European experience. *Baltic Region, 13*(4), 4–26. DOI: 10.5922/2079-8555-2021-4-1

Kyle, P., & Kim, S. H. (2011). Long-term implications of alternative light-duty vehicle technologies for global greenhouse gas emissions and primary energy demands. *Energy Policy, 39*(5), 3012–3024. https://www.sciencedirect.com/science/article/abs/pii/S0301421511001960. DOI: 10.1016/j.enpol.2011.03.016

Kyriakou, D. G., Kanellos, F. D., & Ipsakis, D. (2024). Multi-agent-based real-time operation of microgrids employing plug-in Electric Vehicles and building prosumers. Sustainable Energy, Grids and Networks, 37, 101229. https://www.sciencedirect.com/science/article/pii/S2352467723002370

Lammi, M., & Pantzar, M. (2019). The data economy: How technological change has shaped the citizen-consumer. *Technology in Society, 59*, 101157. Advance online publication. DOI: 10.1016/j.techsoc.2019.101157

Laporte, S., Coquery, G., Deniau, V., De Bernardinis, A., & Hautière, N. (2019). Dynamic wireless power transfer charging infrastructure for future EVs: From experimental track to real circulated roads demonstrations. *World Electric Vehicle Journal, 10*(4), 84. Advance online publication. DOI: 10.3390/wevj10040084

Lazutkina., ... (2019). Lazutkina, V.S., Pokusaev, O.N., Kupriyanovsky, V.P., Sinyakov, S.A. Economic effects of autonomous (unmanned) cars. *International Journal of Open Information Technologies*, 2: 1-15. URL: https://cyberleninka.ru/article/n/ekonomicheskie-effekty-avtonomnyh-bespilotnyh-avtomobiley (date of request: 12/21/2024). (In Russ.)

Lee, Y. J., Khaligh, A., & Emadi, A. (2009). Advanced integrated bidirectional AC/DC and DC/DC converter for plug-in hybrid Electric Vehicles. *IEEE Transactions on Vehicular Technology, 58*(8), 3970–3980. https://ieeexplore.ieee.org/abstract/document/5170004. DOI: 10.1109/TVT.2009.2028070

Lin, F. Y., Carretero, C., Covic, G. A., & Boys, J. T. (2016). A reduced order model to determine the coupling factor between magnetic pads used in wireless power transfer. *IEEE Transactions on Transportation Electrification, 3*(2), 321–331. DOI: 10.1109/TTE.2016.2628793

Lin, F. Y., Covic, G. A., & Kesler, M. (2020). Design of a SAE compliant multicoil ground assembly. *IEEE Journal of Emerging and Selected Topics in Industrial Electronics, 1*(1), 14–25. DOI: 10.1109/JESTIE.2020.2999597

Lin, F. Y., Zaheer, A., Budhia, M., & Covic, G. A. (2014). *Reducing leakage flux in IPT systems by modifying pad ferrite structures. In 2014 IEEE Energy Conversion Congress and Exposition (ECCE).* IEEE., DOI: 10.1109/ECCE.2014.6953630

Lin, F. Y., Zaheer, A., Budhia, M., & Covic, G. A. (2015). Evaluation of magnetic pad sizes and topologies for electric vehicle charging. *IEEE Transactions on Power Electronics, 30*(11), 6391–6407. DOI: 10.1109/TPEL.2015.2419592

Litman, T. (2009). Sustainable transportation indicators: A recommended research program for developing sustainable transportation indicators and data. Paper presented at the 88th Transportation Research Board (TRB) Annual Meeting, Washington, DC.

Litman, T. (2007). Developing indicators for comprehensive and sustainable transport planning. *Transportation Research Record: Journal of the Transportation Research Board, 2017*(1), 10–15. DOI: 10.3141/2017-02

Litman, T. (2016). *Evaluating Public Transportation Health Benefits.* Victoria Transport Policy Institute.

Litman, T., & Burwell, D. (2006). Issues in sustainable transportation. *International Journal of Global Environmental Issues, 6*(4), 331–347. DOI: 10.1504/IJGENVI.2006.010889

Liu, N., & Habetler, T. G. (2013). A study of designing a universal inductive charger for electric vehicles. In IECON 2013—39th Annual Conference of the IEEE Industrial Electronics Society (pp. 4528–4533). IEEE. DOI: 10.1109/IECON.2013.6699865

Liu, Y., Hui, F., Xu, R., Chen, T., Xu, X., & Li, J. (2011). Investigation on the Construction Mode of the Charging Station and Battery-Exchange Station. Asia-Pacific Power and Energy Engineering Conference, APPEEC. 1-2. . https://ieeexplore.ieee.org/abstract/document/5777909DOI: 10.1109/APPEEC.2011.5748626

Liu, C., Jiang, C., & Qiu, C. (2017). Overview of coil designs for wireless charging of electric vehicle. In *2017 IEEE PELS Workshop on Emerging Technologies: Wireless Power Transfer (WoW)* (pp. 1–6). IEEE. DOI: 10.1109/WoW.2017.7959389

Live Streets. (2013, October). Пересадки или прямые маршруты? [Transfers or direct routes?]. http://www.livestreets.ru/2013/10/peresadki-ili-pryamye-marshruty/

Li, X., Wang, Y., & Zhang, P. (2021). Integrating remote sensing and GIS for road damage detection in post-disaster scenarios. *Natural Hazards*, *107*(3), 2029–2047. DOI: 10.1007/s11069-021-04613-0

Liyanage, C., Dias, N., Amaratunga, D., & Haigh, R. P. (2017). Current Context of Transport Sector in South Asia: Recommendations towards a Sustainable System. *Built Environment Project and Asset Management*, *7*(5), 490–505. DOI: 10.1108/BEPAM-10-2016-0051

Luchko, M. I., Tsarev, M. N., & Kovtun, N. V. (2020). Анализ каршеринговых сервисов - преимущества и недостатки [Analysis of carsharing services: Advantages and disadvantages]. *Bulletin of the Volodymyr Dahl Luhansk National University*, (7), 128–131.

Madawala, U. K., Neath, M., & Thrimawithana, D. J. (2013). A power–frequency controller for bidirectional inductive power transfer systems. *IEEE Transactions on Industrial Electronics*, *60*(1), 310–317. DOI: 10.1109/TIE.2011.2174537

Maeda, H., Sekimoto, Y., & Kashiyama, T. (2018). Road damage detection using deep neural networks with images captured through a smartphone. *Computer-Aided Civil and Infrastructure Engineering*, *33*(12), 1127–1141. DOI: 10.1111/mice.12387

Manoharan, S., Mahalakshmi, B., Ananthi, K., & Elakya, A. (2024, March). A Review on Smartphone Charger: Technologies and Challenges. In 2024 5th International Conference on Intelligent Communication Technologies and Virtual Mobile Networks (ICICV) (pp. 84-88). IEEE. https://ieeexplore.ieee.org/abstract/document/10511183

Martus, E. (2020). Municipal solid waste management in Russia: Protest, policy, and politics. *Post-Soviet Affairs*, *36*(5-6), 453–474. DOI: 10.1080/1060586X.2020.1793432

May, A. D., Page, M., & Hull, A. (2001). Developing a set of decision-support tools for sustainable urban transport in the UK. *Transport Policy*, *15*(6), 328–340. DOI: 10.1016/j.tranpol.2008.12.010

Mbusi, A. J. (2024). Reimagining Electric Vehicles for a Sustainable Future. https://www.preprints.org/manuscript/202406.1200/v1

McGarry, C., Dixon, J., Flower, J., Bukhsh, W., Brand, C., Bell, K., & Galloway, S. (2024). Electrified heat and transport: Energy demand futures, their impacts on power networks and what it means for system flexibility. *Applied Energy*, *360*, 122836. https://www.sciencedirect.com/science/article/pii/S0306261924002198. DOI: 10.1016/j.apenergy.2024.122836

Miller, K. (2020). The Tripple Bottom Line: What it is & Why it is Important. *HBS Online*. Accessed via: https://online.hbs.edu/blog/post/what-is-the-triple-bottom-line

Miller, P. (2014). *Sustainability and public transportation: Theory and analysis* (Doctoral dissertation). Calgary, Manitoba: University of Calgary.

Miller, P., de Barros, A. G., Kattan, L., & Wirasinghe, S. C. (2016). Analyzing the sustainability performance of public transit. *Transportation Research Part D, Transport and Environment*, *44*, 177–198. DOI: 10.1016/j.trd.2016.02.012

Miskolczi, M., Földes, D., Munkácsy, A., & Jászberényi, M. (2021). Urban mobility scenarios until the 2030s. *Sustainable Cities and Society*, *72*, 103029. DOI: 10.1016/j.scs.2021.103029

Mohamed, A. A. S., Berzoy, A., de Almeida, F. G. N., & Mohammed, O. (2017). Modeling and assessment analysis of various compensation topologies in bidirectional IWPT system for EV applications. *IEEE Transactions on Industry Applications*, *53*(5), 4973–4984. DOI: 10.1109/TIA.2017.2700281

Mohamed, A. A. S., Shaier, A. A., Metwally, H., & Selem, S. I. (2020). A comprehensive overview of inductive pad in electric vehicles stationary charging. *Applied Energy*, *262*, 114584. Advance online publication. DOI: 10.1016/j.apenergy.2020.114584

Mohammed, A., Saif, O., Abo-Adma, M., Fahmy, A., & Elazab, R. (2024). Strategies and sustainability in fast charging station deployment for Electric Vehicles. *Scientific Reports*, *14*(1), 283. https://www.nature.com/articles/s41598-023-50825-7. DOI: 10.1038/s41598-023-50825-7 PMID: 38168937

Mojumder, M. R. H., Ahmed Antara, F., Hasanuzzaman, M., Alamri, B., & Alsharef, M. (2022). Electric vehicle-to-grid (V2G) technologies: Impact on the power grid and battery. *Sustainability (Basel)*, *14*(21), 13856. Advance online publication. DOI: 10.3390/su142113856

Mosammam, A. M., Rasekh, N., Mirsalim, M., & Moghani, J. S. (2019). *Comparative analysis of the conventional magnetic structure pads for the wireless power transfer applications. In 2019 10th International Power Electronics, Drive Systems and Technologies Conference (PEDSTC)*. IEEE., https://emtrl.aut.ac.ir/publication/conference/

Mou, X., Gladwin, D. T., Zhao, R., Sun, H., & Yang, Z. (2020). Coil design for wireless vehicle-to-vehicle charging systems. *IEEE Access : Practical Innovations, Open Solutions*, *8*, 172723–172733. DOI: 10.1109/ACCESS.2020.3025787

Mrityunjaya, D. H., & Kumar, N., Laxmikant, Ali, S., & Kelagadi, H. M. (2017). Smart transportation. In *2017 International Conference on I-SMAC (IoT in Social, Mobile, Analytics and Cloud) (I-SMAC)* (pp. 1–5). IEEE. DOI: 10.1109/I-SMAC.2017.8058235

Musavi, F., & Eberle, W. (2014). Overview of wireless power transfer technologies for electric vehicle battery charging. *IET Power Electronics*, *7*(1), 60–66. DOI: 10.1049/iet-pel.2013.0047

Mutiawati, C., Suryani, F. M., Anggraini, R., & Isya, M. (2022). Importance-Performance Analysis in Public Transport Level of Service: A Case Study of the Tran Koetaradja Bus in Banda Aceh. *Aceh International Journal of Science and Technology (Banda Aceh)*, *11*(1), 70–84. DOI: 10.13170/aijst.11.1.23146

Naim, N. H., Adnan, M. S. G., Dewan, A., & Zannat, K. E. (2022). Assessing the Performance of Public Transport Services in a Developing Country: A Case Study Using Data Envelopment Analysis. *Growth and Change*, *53*(1), 377–409. DOI: 10.1111/grow.12588

Nasrollahi, Z., Hashemi, M., Bameri, S., & Taghvaee, V. M. (2018). Environmental pollution, economic growth, population, industrialization, and technology in weak and strong sustainability: Using STIRPAT model. *Environment, Development and Sustainability*, *22*(2), 1105–1122. DOI: 10.1007/s10668-018-0237-5

Neustroeva, A., & Shishigina, A. (2022). Food supply in the Arctic: Municipal authorities' opinion. *RUDN Journal of Sociology*, *22*(3), 661–671. DOI: 10.22363/2313-2272-2022-22-3-661-671

NGroom. (2013) Electric Car Maker Tesla Unveils 90s Battery Pack Swap. https://www.reuters.com/ article/2013/06/21/ us-tesla-swap-idUSBRE95K07H20130621

Nguyen, D. M., Kishk, M. A., & Alouini, M. S. (2024). Dynamic charging as a complementary approach in modern EV charging infrastructure. *Scientific Reports*, *14*(1), 5785. https://www.nature.com/articles/s41598-024-55863-3. DOI: 10.1038/s41598-024-55863-3 PMID: 38461152

Nikanorova, L. V., & Loseva, M. V. (2019). Краткосрочная аренда автомобилей - каршеринг [Short-term car rental: Carsharing]. *Modern Technologies and Scientific and Technical Progress*, *1*, 198–199. DOI: 10.36629/2686-9896/2019-1-1-198-199

Nikolaou, I. E., & Tsalis, T. (2018). A framework to evaluate eco- and social-labels for designing a sustainability consumption label to measure strong sustainability impact of firms/products. *Journal of Cleaner Production, 182*, 105–113. DOI: 10.1016/j.jclepro.2018.02.042

Nutkani, I., Toole, H., Fernando, N., & Andrew, L. P. C. (2024). Impact of EV charging on electrical distribution network and mitigating solutions–A review. IET Smart Grid. https://ietresearch.onlinelibrary.wiley.com/doi/full/10.1049/stg2.12156

OECD. (2002). *OECD Guidelines Towards Environmentally Sustainable Transport*. Organization for Economic Co-operation and Development.

Oraegbune, M. O. (2015). Appraisal of Transportation Infrastructure Sustainability in Nigeria.(Ph.D. Thesis, Department of Civil Engineering, Faculty of Engineering, University of Nigeria, Nsukka).

Oraegbune, O. M. (2022). Towards Sustainable Road Infrastructure Delivery in Nigeria. *Arid Zone Journal of Engineering. Technology & Environment, 18*(2), 319–330.

Oza-Frank, R., Warnock, A. L., Calancie, L., Bassarab, K., Palmer, A., Cooksey Stowers, K., & Harris, D. (2025). Food policy councils and healthy food access policies: A 2021 national survey of community policy supports. *Preventing Chronic Disease, 22*, 240335. DOI: 10.5888/pcd22.240335 PMID: 39784113

Panchal, C., Stegen, S., & Lu, J. (2018). Review of static and dynamic wireless electric vehicle charging system. *Engineering Science and Technology, an International Journal, 21*(5), 922–937. DOI: 10.1016/j.jestch.2018.06.015

Pan, S., Yu, W., Fulton, L. M., Jung, J., Choi, Y., & Gao, H. O. (2023). Impacts of the large-scale use of passenger Electric Vehicles on public health in 30 US. metropolitan areas. *Renewable & Sustainable Energy Reviews, 173*, 113100. https://www.sciencedirect.com/science/article/abs/pii/S1364032122009819. DOI: 10.1016/j.rser.2022.113100

Patil, D., McDonough, M. K., Miller, J. M., Fahimi, B., & Balsara, P. T. (2018). Wireless power transfer for vehicular applications: Overview and challenges. *IEEE Transactions on Transportation Electrification, 4*(1), 3–37. DOI: 10.1109/TTE.2017.2780627

Pelegov, D. V., & Chanaron, J. (2022). Electric Car Market Analysis Using Open Data: Sales, Volatility Assessment, and Forecasting. Sustainability. https://www.mdpi.com/2071-1050/15/1/399

Petersen, M., & Fuchs, F. W. (2014). Development of a 5 kW inductive power transfer system including control strategy for electric vehicles. In 2014 International Exhibition and Conference for Power Electronics, Intelligent Motion, Renewable Energy and Energy Management (PCIM) (pp. 1–8). IEEE. DOI: 10.1109/JPROC.2013.2244536

Petrov, A., Zakharov, D., Evtukov, S., Grakov, F., & Petrova, D. (2022). Evaluation of the orderliness of the road safety provision systems in Russian cities. *Frontiers in Built Environment*, *8*, 981078. DOI: 10.3389/fbuil.2022.981078

Pettigrew, S., Booth, L., Farrar, V., Brown, J., Godic, B., & Thompson, J. (2024). An emerging food policy domain: The effects of autonomous transport technologies on food access and consumption. *Food Policy*, *102647*, 102647. Advance online publication. DOI: 10.1016/j.foodpol.2024.102647

Pevere, A., Petrella, R., Mi, C. C., & Zhou, S. (2014). Design of a high efficiency 22 kW wireless power transfer system for EVs fast contactless charging stations. In *2014 IEEE International Electric Vehicle Conference (IEVC)* (pp. 1–7). IEEE. DOI: 10.1109/IEVC.2014.7056221

Podgaiskaya, A. E. (2019). Факторы успеха каршеринговых сервисов в России [Factors of success of carsharing services in Russia]. Region. *Economia e Sociologia*, (2), 270–289. DOI: 10.15372/REG20190212

Poleshkina, I. (2018). Problems of food security in the regions of the Far North of Russia. *Economy of Region*, *14*(3), 820–835. DOI: 10.17059/2018-3-10

Prillinger, H. (n.d.). Tarife [Fares]. University of Vienna. https://homepage.univie.ac.at/horst.prillinger/ubahn/deutsch/tarife.html

Purnama, C., & Subroto, W. T. (2016). Competition Intensity, Uncertainty Environmental on the use of Information Technology and its Impact on Business Performance of Small and Medium Enterprises. *International Review of Management and Marketing*, *6*(4), 984–992.

Pustokhina, I., Lyubko, K. S., & Aleksandrova, A. S. (2021). Autonomous shipping in Russia. *Lizing (Leasing)*, *8*(4), 39–42. DOI: 10.33920/VNE-03-2108-04

Pyankova, S., & Ergunova, O. (2018). The role of regional development institutions in socio-economic development. In SGEM2018 Conference Proceedings: 5th International Multidisciplinary Scientific Conference on Social Sciences and Art (Vol. 5, Issue 1.3, pp. 885–892). SGEM. https://www.sgemsocial.org/ssgemlib/spip.php?article6063&lang=en

Qadir, S. A., Ahmad, F., Al-Wahedi, A. M. A., Iqbal, A., & Ali, A. (2024). Navigating the complex realities of electric vehicle adoption: A comprehensive study of government strategies, policies, and incentives. Energy Strategy Reviews, 53, 101379. https://www.sciencedirect.com/science/article/pii/S2211467X24000865

Qin, X., Ke, J., Wang, X., Tang, Y., & Yang, H. (2022). Demand management for smart transportation: A review. *Multimodal Transportation*, *1*(4), 100038. DOI: 10.1016/j.multra.2022.100038

Rambler. (2024, August 4). Пассажиропоток наземного транспорта Москвы вырос на 6 млн человек в марте [Passenger traffic of Moscow's ground transport increased by 6 million people in March]. https://news.rambler.ru/moscow_city/525 41549passazhiropotok-nazemnogo-transporta-moskvy-vyros-na-6-mln-chelovek-v -marte/

Ramezani, A., & Narimani, M. (2021). A modular coil design for high-power wireless EV charging. In IECON 2021—47th Annual Conference of the IEEE Industrial Electronics Society (pp. 1–5). IEEE. DOI: 10.1109/IECON48115.2021.9589747

Rawson, M., & Kateley, S. (1999). Electric vehicle charging equipment design and health and safety codes. *SAE Transactions*, •••, 3256–3262. https://www.jstor.org/stable/44733996. DOI: 10.4271/1999-01-2941

Reddy, B. S., & Balachandra, P. (2012). Urban mobility: A comparative analysis of megacities of India. *Transport Policy*, *21*, 152–164. DOI: 10.1016/j.tranpol.2012.02.002

Report, B. (1987). Report of the World Commission on Environment and Development: Our Common Future. *United Nations*. Retrieved from https://sustainabledevelopment.un.org/content/documents/5987our-common-future

Rietmann, N., Hügler, B., & Lieven, T. (2020). Forecasting the trajectory of electric vehicle sales and the consequences for worldwide CO2 emissions. *Journal of Cleaner Production*, *261*, 121038. https://www.sciencedirect.com/science/article/abs/pii/S0959652620310854. DOI: 10.1016/j.jclepro.2020.121038

Rosu, S. G., Khalilian, M., Cirimele, V., & Guglielmi, P. (2016). A dynamic wireless charging system for electric vehicles based on DC/AC converters with SiC MOSFET-IGBT switches and resonant gate-drive. In 2016 IEEE 42nd Annual Conference of the IEEE Industrial Electronics Society (IECON) (pp. 4465–4470). IEEE. DOI: 10.1109/IECON.2016.7793809

Ryzhkov, A. (2018). Local public transport in Russia: Regulation, ownership and competition. *Research in Transportation Economics*, *69*, 207–216. DOI: 10.1016/j.retrec.2018.04.010

Ryzhkov, A., & Sarzhan, Y. (2020). Market initiative and central planning: A study of the Moscow bus network. *Research in Transportation Economics*, *83*, 100919. DOI: 10.1016/j.retrec.2020.100919

Sachs, J. D. (2015). *The Age of Sustainable Development*. Columbia University Press. DOI: 10.7312/sach17314

Safiullin, I., Ziganshin, B., Amirova, E., Klychova, G., & Nizamutdinov, M. (2021). Assessment of food security in Russia. *Vestnik of Kazan State Agrarian University*, *16*(3), 124–132. DOI: 10.12737/2073-0462-2021-124-132

Saxena, A., & Gupta, V. (2023). Carpooling: Who is Closest to Adopting it? An Investigation into the Potential Car-poolers among Private Vehicle Users: A Case of a Developing Country, India. *Transport Policy*, *135*, 11–20. DOI: 10.1016/j.tranpol.2023.03.007

Schipper, F., Emanuel, M., & Oldenziel, R. (2020). Sustainable urban mobility in the present, past, and future. *Technology and Culture*, *61*(1), 307–317. DOI: 10.1353/tech.2020.0004 PMID: 32249226

Semyachkov, K. A. (2025). Combined mechanism of public management of a smart city in the digitalization of ecosystem interactions (Doctoral dissertation). Yekaterinburg.

Sergeant, P., & Van den Bossche, A. (2008). Inductive coupler for contactless power transmission. *IET Electric Power Applications*, *2*(1), 1–7. https://digital-library.theiet.org/content/journals/10.1049/iet-epa_20070059. DOI: 10.1049/iet-epa:20070059

Serrano, W. (2018). Digital systems in smart city and infrastructure: Digital as a service. *Smart Cities*, *1*(1), 134–154. DOI: 10.3390/smartcities1010008

Sethuraman, R., & Rudhramoorthy, M. (2024). Performance of Bidirectional ON-Board Charger in Electric Vehicle: A Review. e-Prime-Advances in Electrical Engineering, Electronics and Energy, 100613. https://www.sciencedirect.com/science/article/pii/S2772671124001931

Shagaida, N. I., Uzun, V. Y., Ternovskiy, D. S., & Shishkina, E. A. (2024). Assessment of the economic availability of food in Russia in the context of food security. *Voprosy Ekonomiki*, *6*(6), 73–95. DOI: 10.32609/0042-8736-2024-6-73-95

Shahed, M. T., & Rashid, A. H. U. (2024). Battery charging technologies and standards for Electric Vehicles: A state-of-the-art review, challenges, and future research prospects. *Energy Reports*, *11*, 5978–5998. https://www.sciencedirect.com/science/article/pii/S2352484724003408. DOI: 10.1016/j.egyr.2024.05.062

Shaukat, N., Khan, B., Ali, S. M., Mehmood, C. A., Khan, J., Farid, U., Majid, M., Anwar, S. M., Jawad, M., & Ullah, Z. (2018). A survey on electric vehicle transportation within smart grid system. *Renewable & Sustainable Energy Reviews*, *81*, 1329–1349. https://www.sciencedirect.com/science/article/abs/pii/S1364032117307190. DOI: 10.1016/j.rser.2017.05.092

Shcherba, V., Baranova, K., Abramova, E., Mazaev, A., & Vorobyev, K. (2023). Features of legal regulation and management of municipal solid waste in Russia and Germany. Russian Journal of Resources. *Conservation & Recycling*, *10*(2), 23. DOI: 10.15862/12NZOR223

She, S., Lu, Q., & Ma, C. (2012). A probability–time&space trade-off model in environmental risk perception. *Journal of Risk Research*, *15*(2), 223–234. DOI: 10.1080/13669877.2011.634515

Shimizu, R., Kaneko, Y., & Abe, S. (2013). A new HE core transmitter of a contactless power transfer system that is compatible with circular core receivers and H-shaped core receivers. In 2013 3rd International Electric Drives Production Conference (EDPC) (pp. 1–7). IEEE. DOI: 10.3390/wevj10030049

Sibbing, L., & Candel, J. J. L. (2020). Realizing urban food policy: A discursive institutionalist analysis of Ede municipality. *Food Security*, *12*, 1311–1319. DOI: 10.1007/s12571-020-01126-8

Sibbing, L., Candel, J. J. L., & Termeer, K. (2019). A comparative assessment of local municipal food policy integration in the Netherlands. *International Planning Studies*, *24*(3-4), 241–254. DOI: 10.1080/13563475.2019.1674642

Singh, A., Gurtu, A., & Singh, R. K. (2020). Selection of Sustainable Transport Systems: A case study. *Management of Environmental Quality*, *32*(1), 100–113. DOI: 10.1108/MEQ-03-2020-0059

Sinha, S., Kumar, A., Regensburger, B., & Afridi, K. K. (2019). A new design approach to mitigating the effect of parasitics in capacitive wireless power transfer systems for electric vehicle charging. *IEEE Transactions on Transportation Electrification*, *5*(4), 1040–1059. DOI: 10.1109/TTE.2019.2931869

Spangenberg, J. (2005). Economic sustainability of the economy: Constructs and indicators. *International Journal of Sustainable Development*, *8*(1/2), 47–64. DOI: 10.1504/IJSD.2005.007374

Srivastava, A., Manas, M., & Dubey, R. K. (2023). Electric vehicle integration's impacts on power quality in distribution network and associated mitigation measures: A review. *Journal of Engineering and Applied Sciences (Asian Research Publishing Network)*, *70*(1), 32. https://link.springer.com/article/10.1186/s44147-023-00193-w

Stefansson, G., & Lumsden, K. (2009). Performance issues of smart transportation management systems. *International Journal of Productivity and Performance Management*, *58*(1), 55–70. DOI: 10.1108/17410400910921083

Sun, X., Li, Z., Wang, X., & Li, C. (2020). Technology Development of Electric Vehicles: A Review. *Energies*, *90*. Advance online publication. https://www.mdpi.com/1996-1073/13/1/90. DOI: 10.3390/en13010090

Susarenko, M. Y. (2023). Car sharing as an integral element of the development of the electric vehicle transport market. *The Scientific Heritage*, (122), 17–19. DOI: 10.5281/zenodo.8412786

Su, W., Eichi, H., Zeng, W., & Chow, M. Y. (2011). A survey on the electrification of transportation in a smart grid environment. *IEEE Transactions on Industrial Informatics*, *8*(1), 1–10. https://ieeexplore.ieee.org/abstract/document/6051485. DOI: 10.1109/TII.2011.2172454

Taghvaee, M. V., Arani, A. A., Nodehi, M., Shirazi, J. K., Agheli, L., Ghojogh, H. M. N., Salehnia, N., Mirzaee, A., Taheri, S., Saber, R. M., Faramarzi, H., Alvandi, R., & Rahbarian, H. A. (2023).. . *Review of Economics and Political Science*, *8*(2), 134–161. DOI: 10.1108/REPS-12-2019-0168

Takahashi, R., Negishi, K., Noda, H., & Mizutani, M. (2023). Estimating the Dominant Life Phase Concerning the Effects of Battery Degradation on CO2 Emissions by Repetitive Cycle Applications: Case Study of an Industrial Battery System Installed in an Electric Bus. *Energies*, *16*(3), 1508. https://www.mdpi.com/1996-1073/16/3/1508. DOI: 10.3390/en16031508

Tang, Y., Zhu, F., Wang, Y., & Ma, H. (2015). Design and optimizations of solenoid magnetic structure for inductive power transfer in EV applications. In IECON 2015—41st Annual Conference of the IEEE Industrial Electronics Society (pp. 001459–001464). IEEE. DOI: 10.1109/IECON.2015.7392306

Tan, L., Zhao, W., Liu, H., Li, J., & Huang, X. (2020). Design and optimization of ground-side power transmitting coil parameters for EV dynamic wireless charging system. *IEEE Access : Practical Innovations, Open Solutions*, *8*, 74595–74604. DOI: 10.1109/ACCESS.2020.2988622

Tejeda, C., Carretero, C., Boys, J. T., & Covic, G. A. (2016). Ferrite-less circular pad with controlled flux cancellation for EV wireless charging. *IEEE Transactions on Power Electronics*, *32*(11), 8349–8359. DOI: 10.1109/TPEL.2016.2642192

The Maritime Executive. (2020). Transport uses 25 percent of world energy. Retrieved from https://www.maritime-executive.com/article/transport-uses-25-percent-of-world-energy

Throngnumchai, K., Kai, T., & Minagawa, Y. (2011, September). A study on receiver circuit topology of a cordless battery charger for Electric Vehicles. In 2011 IEEE Energy Conversion Congress and Exposition (pp. 843-850). IEEE. https://ieeexplore.ieee.org/abstract/document/6063858

Tian, X., & Lu, H. (2023). Digital infrastructure and cross-regional collaborative innovation in enterprises. *Finance Research Letters*, *58*(C), 104635. DOI: 10.1016/j.frl.2023.104635

Tintelecan, A., Dobra, A. C., & Marţiş, C. (2019, October). LCA indicators in Electric Vehicles environmental impact assessment. In 2019 Electric Vehicles International Conference (EV) (pp. 1-5). IEEE. https://ieeexplore.ieee.org/abstract/document/8892893

Tuttle, D. P., & Baldick, R. (2012). The evolution of plug-in electric vehicle-grid interactions. *IEEE Transactions on Smart Grid*, *3*(1), 500–505. DOI: 10.1109/TSG.2011.2168430

Uldanov, A. (2019). Policy advice in an authoritarian environment: Urban transport policies in Moscow and Beijing (2010–2017). *Policy Studies*, *40*(1), 58–79. DOI: 10.1080/01442872.2019.1581158

UN-Habitat. (2020). *International guidelines on urban and territorial planning*. https://unhabitat.org/sites/default/files/2020/04/ig-utp_russian.pdf

Unified Interdepartmental Information and Statistical System. (n.d.). Unified Interdepartmental Information and Statistical System. Retrieved February 7, 2025, from https://www.fedstat.ru/indicator/58939

United Nations. (2021). Sustainable transport, sustainable development. *Interagency report for the Second Global Sustainable Transport Conference*.

Un-Noor, F., Padmanaban, S., Mihet-Popa, L., Mollah, M. N., & Hossain, E. (2017). A comprehensive study of key electric vehicle (EV) components, technologies, challenges, impacts, and future direction of development. *Energies*, *10*(8), 1217. Advance online publication. DOI: 10.3390/en10081217

Van de Graaf, T., & Lesage, D. (2009). The International Energy Agency after 35 years: Reform needs and institutional adaptability. *The Review of International Organizations*, *4*(3), 293–317. DOI: 10.1007/s11558-009-9063-8

Van Wee, B., & Handy, S. (2015). Key research themes on urban space, scale, and sustainable urban mobility. *International Journal of Sustainable Transportation*, *10*(1), 18–24. DOI: 10.1080/15568318.2013.820998

vanBurgsteden, M., Grigolon, A., & Geurs, K. T. (2024). Improving community wellbeing through transport policy: A literature review and theoretical framework, based on the capability approach. *Transport Reviews*, *44*(2), 297–320. DOI: 10.1080/01441647.2024.2374818

Verma, B., Snodgrass, R., Henry, B., Smith, B., & Daim, T. (2020). Smart cities—An analysis of smart transportation management. In Tiwari, R., & Buse, S. (Eds.), *Managing innovation in a global and digital world* (pp. 401–422). Springer Gabler., DOI: 10.1007/978-3-658-27241-8_23

Vicente, P., Sampaio, A., & Reis, E. (2020). Factors influencing passenger loyalty towards public transport services: Does public transport providers' commitment to environmental sustainability matter? *Case Studies on Transport Policy*, *8*(2), 627–638. DOI: 10.1016/j.cstp.2020.02.004

Villa, J. L., Sallán, J., Llombart, A., & Sanz, J. F. (2009). Design of a high frequency inductively coupled power transfer system for electric vehicle battery charge. *Applied Energy*, *86*(3), 355–363. DOI: 10.1016/j.apenergy.2008.05.009

Wang, L., & Shao, J. (2024). The energy saving effects of digital infrastructure construction: Empirical evidence from Chinese industry. *Energy*, *294*, 130778. DOI: 10.1016/j.energy.2024.130778

Wang, X., Chan, H. K., Yee, R. W., & Diaz-Rainey, I. (2012). A two-stage fuzzy-AHP model for risk assessment of implementing green initiatives in the fashion supply chain. *International Journal of Production Economics*, *135*(2), 595–606. DOI: 10.1016/j.ijpe.2011.03.021

Wegren, S., Nilssen, F., & Elvestad, C. (2016). The impact of Russian food security policy on the performance of the food system. *Eurasian Geography and Economics*, *57*(6), 671–699. DOI: 10.1080/15387216.2016.1222299

Weicker, T. (2020). Marshrutka (in)formality in southern Russian cities and its role in contentious transport policies. *Geoforum*, *114*, 159–168. DOI: 10.1016/j.geoforum.2020.08.015

Weicker, T. (2022). Transport reforms and its missing publics: Insights from marshrutka abolishment and transport 'modernisation' policies prior to FIFA World Cup 2018 in Volgograd, Russian Federation. *Urban Studies (Edinburgh, Scotland)*, *59*(12), 2531–2547. DOI: 10.1177/00420980221135405

Xu, H., Lin, J., & Yu, W. (2017). Smart transportation systems: Architecture, enabling technologies, and open issues. In Sun, Y., & Song, H. (Eds.), *Secure and trustworthy transportation cyber-physical systems* (pp. 13–26). Springer., DOI: 10.1007/978-981-10-3892-1_2

Yamada, Y., Sasaki, K., Imura, T., & Hori, Y. (2021). Design method of coils for dynamic wireless power transfer considering average transmission power and installation rate. In *2021 IEEE Southern Power Electronics Conference (SPEC)* (pp. 1–8). IEEE. DOI: 10.1109/SPEC52827.2021.9709485

Yang, G., Song, K., Huang, X., Wang, C., Huang, X., Li, J., & Zhang, Y. (2021a). Improved interoperability evaluation method for wireless charging systems based on interface impedance. *IEEE Transactions on Power Electronics*, *36*(8), 8588–8592. DOI: 10.1109/TPEL.2021.3053353

Yang, G., Song, K., Sun, Y., Huang, X., Li, J., Guo, Y., Zhang, Y., Zhang, Q., Lu, R., & Zhu, C. (2021b). Interoperability improvement for rectangular pad and DD pad of wireless electric vehicle charging system based on adaptive position adjustment. *IEEE Transactions on Industry Applications*, *57*(3), 2613–2624. DOI: 10.1109/TIA.2021.3056639

Yan, J., Liu, J., & Tseng, F. M. (2020). An evaluation system based on the self-organizing system framework of smart cities: A case study of smart transportation systems in China. *Technological Forecasting and Social Change*, *153*, 119371. DOI: 10.1016/j.techfore.2018.07.009

Yilmaz, M., & Krein, P. T. (2012). Review of battery charger topologies, charging power levels, and infrastructure for plug-in electric and hybrid vehicles. *IEEE Transactions on Power Electronics*, *28*(5), 2151–2169. https://ieeexplore.ieee.org/abstract/document/6280677. DOI: 10.1109/TPEL.2012.2212917

Yin, J., Lin, D., Parisini, T., & Hui, S. Y. (2016). Front-end monitoring of the mutual inductance and load resistance in a series-series compensated wireless power transfer system. *IEEE Transactions on Power Electronics*, *31*(10), 7339–7352. DOI: 10.1109/TPEL.2015.2509962

Yuvaraj, T., Devabalaji, K. R., Kumar, J. A., Thanikanti, S. B., & Nwulu, N. (2024). A comprehensive review and analysis of the allocation of electric vehicle charging stations in distribution networks. IEEE Access. https://ieeexplore.ieee.org/abstract/document/10380227

Zaheer, A., Hao, H., Covic, G. A., & Kacprzak, D. (2014). Investigation of multiple decoupled coil primary pad topologies in lumped IPT systems for interoperable electric vehicle charging. *IEEE Transactions on Power Electronics*, *30*(4), 1937–1955. DOI: 10.1109/TPEL.2014.2329693

Zaheer, A., Kacprzak, D., & Covic, G. A. (2012). *A bipolar receiver pad in a lumped IPT system for electric vehicle charging applications. In 2012 IEEE Energy Conversion Congress and Exposition (ECCE).* IEEE., DOI: 10.1109/ECCE.2012.6342811

Zamyatina, N., Goncharov, R., Poturaeva, A., & Pelyasov, A. (2020). The sandwich of Russian space: How different spaces differentiate themes in regional science. *Regional Science Policy & Practice*, *12*(4), 559–577. DOI: 10.1111/rsp3.12272

Zavyalov D. V. The concept of sustainable mobility in modern megacities // MIR (Modernization. Innovation. Development). 2021. Vol. 12. No. 3. pp. 288-305 https://doi.org/ (In Russ.).DOI: 10.18184/2079-4665.2021.12.3.288-305

Zentani, A., Almaktoof, A., & Kahn, M. T. (2024). A Comprehensive Review of Developments in Electric Vehicles Fast Charging Technology. *Applied Sciences (Basel, Switzerland)*, *14*(11), 4728. https://www.mdpi.com/2076-3417/14/11/4728. DOI: 10.3390/app14114728

Zhang, B., Chen, Q., Zhang, L., Chen, J., Xu, L., Ren, X., & Zhang, Z. (2021). Triple-coil-structure-based coil positioning system for wireless EV charger. *IEEE Transactions on Power Electronics*, *36*(12), 13515–13525. DOI: 10.1109/TPEL.2021.3085604

Zhang, W., Mo, Q., Yan, P., & Liu, Y. (2021). Simultaneous metal object detection and coil alignment for wireless EV chargers using planar coil array. *IEEE Transactions on Instrumentation and Measurement*, *70*, 4507411. Advance online publication. DOI: 10.1109/TIM.2021.3126394

Zhang, W., White, J. C., Abraham, A. M., & Mi, C. C. (2015). Loosely coupled transformer structure and interoperability study for EV wireless charging systems. *IEEE Transactions on Power Electronics*, *30*(11), 6356–6367. DOI: 10.1109/TPEL.2015.2433678

Zhang, X., Yuan, Z., Yang, Q., Li, Y., Zhu, J., & Li, Y. (2016). Coil design and efficiency analysis for dynamic wireless charging system for electric vehicles. *IEEE Transactions on Magnetics, 52*(7), 8700404. Advance online publication. DOI: 10.1109/TMAG.2016.2529682

Zhao, C., Dong, K., & Taghizadeh-Hesary, F. (2023). Can smart transportation enhance green development efficiency? *Economic Change and Restructuring, 56*(2), 825–857. DOI: 10.1007/s10644-022-09448-7

Zheng, Y., Dong, Z. Y., Xu, Y., Meng, K., Zhao, J. H., & Qiu, J. (2013). Electric vehicle battery charging/swap stations in distribution systems: Comparison study and optimal planning. *IEEE Transactions on Power Systems, 29*(1), 221–229. https://ieeexplore.ieee.org/abstract/document/6590033. DOI: 10.1109/TPWRS.2013.2278852

Zhou, J. (2012). Sustainable transportation in the US: A review of proposals, policies, and programs since 2000. *Frontiers of Architectural Research, 1*(2), 150–165. DOI: 10.1016/j.foar.2012.02.012

Zhou, X., Lukic, S., Bhattacharya, S., & Huang, A. (2009, September). *Design and control of grid-connected converter in bi-directional battery charger for plug-in hybrid electric vehicle application. In 2009 IEEE Vehicle Power and Propulsion Conference*. IEEE., https://ieeexplore.ieee.org/abstract/document/5289691

Zhou, X., Wang, G., Lukic, S., Bhattacharya, S., & Huang, A. (2009, September). *Multi-function bi-directional battery charger for plug-in hybrid electric vehicle application. In 2009 IEEE energy conversion Congress and exposition*. IEEE., https://ieeexplore.ieee.org/abstract/document/5316226

Zwald, M., Eyler, A., Goins, K. V., Brownson, R., Schmid, T., & Lemon, S. (2017). Understanding municipal officials' involvement in transportation policies supportive of walking and bicycling. *Journal of Public Health Management and Practice, 23*(4), 348–355. DOI: 10.1097/PHH.0000000000000152 PMID: 25319080

About the Contributors

Dinesh Elango is a respected Associate Professor of Business Analytics and Business Management. With a PhD in Management Sciences from Anna University, Chennai, Dr. Dinesh has been recognized as an Accredited Management Teacher (AMT) by the All India Management Association.

Rajesh Verma is a Professor & Sr. Dean at Mittal School of Business (NIRF Ranked), Lovely Professional University (LPU), where he leads the academic and administrative functions of the business school. He has a PhD in Marketing from Himachal Pradesh University, an MBA from Utkal University, and an Executive Programme in Data Driven Product Management from IIM Lucknow. He is also a Wiley Certified Design Thinking Practitioner

***

Sachin Bajirao Bhosale is a dedicated PhD research scholar whose academic journey exemplifies perseverance, commitment, and a passion for knowledge. Currently pursuing his doctoral studies at Sandip University, Nashik, he specializes in Computer Science and Engineering. His research focuses on the critical subject of Deep Learning-Driven Pothole Dimension Analysis for Enhanced Road Maintenance Prioritization: A Comprehensive Framework, a groundbreaking study aimed at addressing key challenges and contributing innovative solutions to the field. He completed his undergraduate and postgraduate studies at Ellenki College of Engineering and Technology, where he consistently demonstrated exceptional academic performance and a deep curiosity for exploring uncharted territories of knowledge. His transition to doctoral research was driven by his unwavering commitment to solving real-world problems through rigorous scientific inquiry. As a scholar, Sachin Bajirao Bhosale is known for his sincerity and unyielding work ethic. He approaches his research with meticulous attention to detail, ensuring that every experiment, analysis, and written contribution reflects the highest standards of academic excellence. His ability to manage complex tasks, coupled with his dedication to continuous learning, has earned him the respect and admiration of his peers and mentors. In addition to his research, Sachin Bajirao Bhosale actively participates in academic conferences, workshops, and seminars, showcasing his findings and engaging with the global academic community. His work not only enriches his discipline but also inspires fellow researchers to pursue excellence in their endeavors.

Delshi Howsalya Devi received her BE (distinction) in Computer Science and Engineering from the Madurai Kamaraj University, Madurai, in 2004, ME in Computer Science and Engineering from the Anna University, Chennai, in 2008 and PhD in Information and Communication Engineering from the Anna University Chennai in 2018. She is a Professor and Head at the Department of Artificial Intelligence and Data Science, Karpaga Vinayaga College of Engineering & Technology, Tamil Nadu. She has 17 years of teaching experience and has approximately 59 conference publications and 27 international journal publications. She has published 27 National Patents and 1 International patent. She reviewed a Research project proposal which was assigned from Sapienza university, Rome. She is associated with IEEE access Journal, Journal of Supercomputing, Springer, Journal of Ambient Intelligence and Humanized Computing, Pattern Recognition Letters Springer, Acta Scientific Pharmacology. She was acted as a technical committee member in many reputed conferences so far. She was received a National Award titled "Young Educator and Research Scholar" by National Foundation for Entrepreneurship Development, and she was acted as panel judge in hackathon. Her research interests include Data mining, Big data analytics, Artificial Intelligence, Data Science, Machine Learning and Deep Learning.

L. Rajesh currently a research scholar in the Department of Mechatronics Engineering at SRM Institute of Science and Technology, Chengalpattu, Chennai. He received a bachelor's degree in engineering in ECE from Sri Ramakrishna Engineering College, Anna University, Coimbatore, in the year 2011. He received his master's degree in Embedded Systems from Kumaraguru College of Technology, Coimbatore, Anna University, in the year 2013. He has more than 8 years of academic experience at various engineering colleges. His current research is focused on wireless power transfer in EVs. He has 8 research publications in national/ international journals and conferences to her credit.

Maiyaki Sheneni Felix is a Research and Teaching Assistant at Lagos Business School, Pan-Atlantic University. He has an academic background in

Entrepreneurship, which has greatly influenced his research focus. His research focuses on innovation for social impact, adopting new technologies, and the sustainability principles essential for modern business practices. He is also the Research Manager at the Lagos Business School Tech-Leap Initiative (LBS-TLI), where he engages in research and industry engagements that drive technological advancement. He has previously held positions related to procurement, inventory management, and operations in the agricultural and FMCG sectors.

Frank Ojadi obtained a bachelor's degree in Engineering Physics from the University of Ife in 1981 (now Obafemi Awolowo University), an MBA from the Maastricht School of Management, Netherlands and PhD in Transport Economics from the Department of Transport and Supply Chain Management of the University of Johannesburg, South Africa. He has been a faculty member at Lagos Business School, Nigeria, for 24 years. He has also provided leadership in the school's DBA programme. Prior to entering academia, he worked at various times as an Operations Planning, Logistics and Supply Chain Manager in the steel, aluminium, paper, pharmaceutical, tobacco and hospitality sub-sectors of the Nigerian economy. Professor Ojadi's teaching and research interests are in operations management, design and implementation of logistics and supply chain management improvement methodologies. His research works have been published in Supply Chain Management: An International Journal; International Journal of Production Research; Journal of Transport and Supply Chain Management; International Journal of Logistics: Research and Applications.

Sivaram Ponnusamy is a distinguished academic and researcher with expertise across diverse fields, including computer science, electrical engineering, and business management. With qualifications such as a B.E. in Electrical and Electronics Engineering, an M.E. in Computer Science, a Ph.D., and an MBA in Project Management, he serves as a professor and research supervisor in engineering colleges and universities, specializing in artificial intelligence and computer science. Dr. Ponnusamy has filed multiple patents in cutting-edge technologies, ranging from remote voting systems to AI-based attendance management and environmental analysis tools. His work includes editing several Scopus-indexed books on topics like AI applications, digital twin technology, and women's safety, with ongoing international projects in AI research. As an active member of various professional bodies, including IAENG and ISOC, he also contributes as a reviewer for reputed journals. Recognized for his academic contributions, he was awarded by the Thanjavur District Collector for developing an innovative Android-based meeting management tool. Additionally, Dr. Ponnusamy frequently chairs international conferences and conducts seminars on the latest technological trends, making significant contributions to the fields of AI, machine learning, and software development.

T. Sangeetha currently, working as Associate professor in the Department of Biomedical Engineering at Karpaga Vinayaga College of Engineering and Technology, Chengalpattu, Chennai. She received Bachelor Degree in Engineering in ECE from Odaiyappa College of Engineering and technology, Anna University, Chennai in the year 2008. She received her Master's Degree in Communication Systems from Thiagarajar College of Engineering, Madurai, and Anna University in the year 2010 and she completed her Ph.D., from Anna University, Chennai, in the year of 2022. She has more than 14 years of academic experience at various engineering college. Her current research focused on Mobile Ad-hoc Networking and her research area is communication and Networking, Signal Processing, Wireless antenna and Wireless Communication, E-Vehicle. She has 17 research publications in National/International Journals and conferences to her credit

Pyankova Svetlana Grigoryevna Usova Natalia Vitalievna has been a member of Dissertation Council, authorized to oversee doctoral and candidate dissertations in the field "Economics and Management of the National Economy" (with specializations in Industry, Regional Economy, and Logistics).S.G. Pyankova is a distinguished scholar whose teaching, research, and practical contributions have been recognized with numerous awards at the university, regional, national, and international levels. In 2017, she was awarded the title of Corresponding Member of the Russian Academy of Natural Sciences and received the honorary title "Distinguished Scientist in Science and Education." She has also been honored with several prestigious awards, including the Adam Smith Medal, the Order of Catherine the Great "For Service to Science and Education," the Order "Labore et Scientia," the Lomonosov Medal, and others. She actively participates in scientific, expert, and outreach activities. She serves on the editorial boards of the academic journals Noonomy and Society and Regional Economics and Management. Since November 2017, she has been a member of the Board and the Presidium of the Ural Branch of the Free Economic Society of Russia, and since 2018, a member of the International Union of Economists. She was also involved in a Russian Foundation for Basic Research (RFBR) grant project on neo-industrialization in the Ural macroregion

Lyu Ye, PhD Associate Professor School of Management, Sichuan University of Science & Engineering Zigong, 643002, China Email: lyuye@suse.edu.cn Dr. Lyu Ye is an Associate Professor at the School of Management, Sichuan University of Science & Engineering, China. She holds a PhD in Economics and specializes in the study of human capital, sustainable development, and innovation management within the context of China's evolving knowledge economy. Her research explores the effectiveness of human capital investment, corporate sustainability, green transformation, and digital strategies in innovative enterprises. Dr. Lyu has authored over 20 scholarly publications in reputable journals indexed by Scopus, Web of Science, and the Russian Higher Attestation Commission (VAK), with a particular emphasis on interdisciplinary collaboration between China and Russia. Her international co-authored works investigate topics such as environmental information disclosure, green credit policy, and the use of deep learning for profitability forecasting in the petroleum sector. She actively contributes to research on structural economic transformation, especially concerning the role of human capital in China's innovation-driven development.

Index

A

B

C

D

E

F

G

H

I

M

P

www.ingramcontent.com/pod-product-compliance
Lightning Source LLC
LaVergne TN
LVHW080855300926
849162LV00003B/426